# Modelling and Process Control of Fuel Cell Systems

# Modelling and Process Control of Fuel Cell Systems

Editors

**Mohd Azlan Hussain**
**Wan Ramli Wan Daud**

MDPI • Basel • Beijing • Wuhan • Barcelona • Belgrade • Manchester • Tokyo • Cluj • Tianjin

*Editors*
Mohd Azlan Hussain
University of Malaya
Malaysia

Wan Ramli Wan Daud
Universiti Kebangsaan Malaysia
Malaysia

*Editorial Office*
MDPI
St. Alban-Anlage 66
4052 Basel, Switzerland

This is a reprint of articles from the Special Issue published online in the open access journal *Processes* (ISSN 2227-9717) (available at: https://www.mdpi.com/journal/processes/special_issues/fuel_cell_model).

For citation purposes, cite each article independently as indicated on the article page online and as indicated below:

LastName, A.A.; LastName, B.B.; LastName, C.C. Article Title. *Journal Name* **Year**, *Volume Number*, Page Range.

**ISBN 978-3-0365-0574-9 (Hbk)**
**ISBN 978-3-0365-0575-6 (PDF)**

# Contents

# About the Editors

**Mohd Azlan Hussain** received his PhD from Imperial College of Technology, University of London. He has been a Professor at the Department of Chemical Engineering, Faculty of Engineering, University of Malaya, for the last 30 years. His activities and expertise have resulted in various outputs, which include the graduation of 24 PhD and 24 Masters students and 7 postdoc researchers; authoring 1 book, 9 book chapters, and over 210 journal papers and 250 conference proceedings papers; being an external examiner for 50 PhD and 38 Masters students locally and abroad; as well as being a paper reviewer for over 120 journals and various books/book chapters. He is also on the editorial board of five international journals related to chemical engineering. The culmination of these achievements resulted in him being awarded the University's Excellent Service Awards in 2006, 2007, 2009, 2011, and 2014, as well as being appointed the Head of Department in 2003, the Deputy Dean of the faculty in 2007, and again the Head of Department in 2014. He was also included as a MARQUIS "Who's Who in the World" personality in 2006 and in the Who's Who in the 21st Century by the International Biographical Centre, Cambridge, in 2001. He won the IChemE Global Award under the Water Category and the Highly Commended Award under the Industry Project in 2018. He was also awarded the Top Research Scientist award by the the Academy of Science, Malaysia, and recently listed in the top 2% of scientists in the world in Chemical Engineering by a study done at Stanford University.

**Wan Ramli Wan Daud** FASc has been the Professor of Chemical Engineering at Department of Chemical & Process Engineering, Faculty of Engineering & Built Environment, Universiti Kebangsaan Malaysia, since 1996. He was formerly the Founding Director of the Fuel Cell Institute, Universiti Kebangsaan Malaysia, in 2007–2014 and a Fellow of the Academy of Sciences Malaysia. He was awarded the BEng degree (First Class Hon.) from the University of Monash in 1978 and the PhD degree from the University of Cambridge in 1984 in Chemical Engineering. He was Head of the same Department in 1985–1988 and the Deputy Dean of Engineering, Universiti Kebangsaan Malaysia, in 1990–1993 and 1995–1998. He is an expert on zero-emission fuel cell technology, hydrogen energy, and sustainable drying technology. He was invited to present 17 international keynote papers, 10 invited international papers and 9 national keynote papers in recognition of his expertise in these three key areas. He has published 839 articles including 275 articles in international journals, 339 articles in the proceedings of international conferences, and 225 articles in the proceedings of national conferences. He has published two international research books and two national books, and contributed five chapters in international research books and five chapters in national books. He was awarded the ASEAN Energy Award 2007, ASEAN Energy Award 2005, the IChemE Highly Commended Shell Energy Award 2008, the Outstanding Contribution to the Drying Community 2009 Award, and the Award for Excellence in Research in Drying of Agricultural Products and Outstanding Contribution to the Development of Drying Technology 2011.

# Preface to "Modelling and Process Control of Fuel Cell Systems"

The ever-increasing energy consumption, rising public awareness for environmental protection, and higher prices of fossil fuels have motivated many to look for alternative and renewable energy sources. The global fossil fluid fuel demand will soon exceed the global fossil fluid fuel production, which is expected to lead to an energy shortage crisis unless a sustainable alternative fuel is available soon. Among the many alternative fuel sources, fuel cells have received a major share of the attention, while they can also act as cogeneration systems. The complicated reaction, heat, and mass transfer mechanisms in the fuel cells introduce extreme nonlinearities in the dynamics of the fuel cell.

The fundamental modeling and control problem in the fuel cells is further complicated by the existence of strong interaction between the input and output parameters; conventional modeling approaches and control strategies are incapable of coping with these difficulties. The conventional models do not consider all these phenomena in their model. Therefore, a comprehensive analysis of the models and effects of various parameters are needed to provide a more realistic understanding of the phenomena encountered in fuel cells and improve the quantitative understanding of the actual process. Since fuel cells are severely nonlinear and typically have several operational constraints, a single linear controller may not provide satisfactory performance over a wide range of operating conditions.

Therefore, advanced process control schemes need to be implemented to cater for the process dynamic nonlinearities and difficulties involved in the robust control of fuel cells. Efficient management and operation of such hybrid fuel cell grids are hence also needed along with better control methods. Since the simulation results of modeling are only predictions and estimations of a real system, an important step in the development of modeling and control is online validation. Unfortunately, there is a lack of experimental validation of the dynamic models of fuel cells in the open literature at present. Environmental assessment of systems closely related to fuel cell operations, such as the lithium-ion battery, is also necessary in these further studies. Hence, this Special Issue aims to highlight the recent trends in these topics, with several papers related to the above issues.

**Mohd Azlan Hussain, Wan Ramli Wan Daud**

*Editors*

*Editorial*

# Special Issue on "Modelling and Process Control of Fuel Cell Systems"

**Mohd Azlan Hussain [1,*] and Wan Ramli Wan Daud [2,*]**

1 Department of Chemical Engineering, Faculty Engineering, University of Malaya, Kuala Lumpur 50603, Malaysia
2 Department of Chemical & Process Engineering, Faculty of Engineering & Built Environment, Universiti Kebangsaan Malaysia, UKM Bangi 43600, Malaysia
* Correspondence: mohd_azlan@um.edu.my (M.A.H.); wramli@ukm.edu.my (W.R.W.D.); Tel.: +60-379-675-214 (M.A.H.); +60-389-118-418 (W.R.W.D.)

Received: 2 December 2020; Accepted: 2 December 2020; Published: 3 December 2020

---

The ever increasing energy consumption, rising public awareness for environmental protection, and higher prices of fossil fuels have motivated many to look for alternative and renewable energy sources. The world fossil fluid fuel demand will soon exceed the world fossil fluid fuel production, which is expected to lead to an energy shortage crisis unless a sustainable alternative fuel is available soon. Among the many alternative fuel sources, fuel cells have received a major share of the attention, while they can also act as cogeneration systems.

The complicated reaction, heat, and mass transfer mechanisms in the fuel cells introduce extreme nonlinearities in the dynamcis of the fuel cell. The fundamental modeling and control problem in the fuel cells is further complicated by the existence of strong interaction between the input and output parameters; conventional modeling approaches and control strategies are incapable of coping with these difficulties. The conventional models do not consider all these the phenomena in their model. Therefore, a comprehensive analysis of the models and effects of various parameters are needed to provide a more realistic understanding of the phenomena encountered in fuel cells and improve the quantitative understanding of the actual process.

Since fuel cells are severely nonlinear and typically have several operational constraints, a single linear controller may not provide satisfactory performance over a wide range of operating conditions. Therefore, advanced process control schemes are needed to be implemented to cater the process dynamic nonlinearities and difficulties involved in the robust control of fuel cells. Efficient management and operation of such hybrid fuel cell grids are hence also needed along with better control methods. Since the simulation results of modeling is only a prediction and estimation of real system, an important step in the development of modeling and control is online validation. Unfortunately, there is a lack of experimental validation of the dynamic models of fuel cells in the open literature at present. Environmental assesment of systems closely related to fuel cell operations such as the Lithium-Ion battery is also necessary in these further studies.

In this special issue, we have seven papers related to the above issues i.e., of Fuel-Cell based Cogeneration System (1 paper), Management and Control of Fuel Cell Systems (2 papers), Analysis, Simulation and Operations of different types of fuel cells (1 paper), Modelling and Online experiment validation (2 papers), and environment assessment of Cathode Materials in Lithium-Ion battery energy generation systems (1 paper).

The paper by Ramadhani, F. et al. [1] gives a comprehensive review with technical guidelines for the design and operation of fuel-cell especially in cogeneration system setup. This review can be an important source of reference for the optimal design and operation of various type of fuel cells in cogeneration systems.

The paper by Atawi, I.E. et al. [2] discusses the modelling, management, and control of an autonomous hybrid microgrid system which incorporates fuel cells. This work utilizes an optimal control algorithm called the Mine Blast Algorithm, where the fuel cell compensates for extra load in the power demands of the system. The paper by Chatrattanawet, N. et al. [3] involves the design and implementation of off line robust model predictive control for solid oxide fuel cells. This work relates to the control of the temperature and fuel in a direct internal reforming solid oxide fuel cell through an ellipsoidal invariant set. For the analysis part, the paper by Ahmed et al. [4] touches on the simulation of solid oxide fuel cell anode using Aspen HYSYS Software. This paper mainly focus on the study of the effect of reforming activity on distributed performance profiles, carbon formation and anode oxidation risks.

At the same time, the paper by Govindarasu, R.; Somasundaram, S. [5] involves the mathematical modelling and simulation to identify the most influencing process variable affecting the fuel cell operation. Real time experiments were carried out to validate and obtain the optimum temperature for maximun power density. Furthermore, the paper by Burkic, D. et al. [6] discusses on the effects of induced friction in open cathode conduits with virtual roughness in the air-forced flow of a proton exchange membrane fuel cell. The regression model obtained correlates air flow and pressure drop as a function of the variable flow friction factor.

Finally, the work of Wang, L. et al. [7] presents the environmental sustainability assessment of typical cathode materials of Lithium-Ion battery based on three life cycle assessment approaches that are applicable to the other cathode-based set up such as the fuel cell systems.

We thank all the contributors as well as the editorial staff of *Processes* for the support of this special issue.

**Funding:** This research received no external funding.

**Conflicts of Interest:** The authors declare no conflict of interest.

## References

1. Ramadhani, F.; Hussain, M.A.; Mokhlis, H. A Comprehensive Review and Technical Guideline for Optimal Design and Operations of Fuel Cell-Based Cogeneration Systems. *Processes* **2019**, *7*, 950. [CrossRef]
2. Atawi, I.E.; Kassem, A.M.; Zaid, S.A. Modeling, Management, and Control of an Autonomous Wind/Fuel Cell Micro-Grid System. *Processes* **2019**, *7*, 85. [CrossRef]
3. Chatrattanawet, N.; Kheawhom, S.; Chen, Y.-S.; Arpornwichanop, A. Design and Implementation of the Off-Line Robust Model Predictive Control for Solid Oxide Fuel Cells. *Processes* **2019**, *7*, 918. [CrossRef]
4. Ahmed, K.; Amiri, A.; Tadé, M.O. Simulation of Solid Oxide Fuel Cell Anode in Aspen HYSYS—A Study on the Effect of Reforming Activity on Distributed Performance Profiles, Carbon Formation, and Anode Oxidation Risk. *Processes* **2020**, *8*, 268. [CrossRef]
5. Govindarasu, R.; Somasundaram, S. Studies on Influence of Cell Temperature in Direct Methanol Fuel Cell Operation. *Processes* **2020**, *8*, 353. [CrossRef]
6. Brkić, D.; Praks, P. Air-Forced Flow in Proton Exchange Membrane Fuel Cells: Calculation of Fan-Induced Friction in Open-Cathode Conduits with Virtual Roughness. *Processes* **2020**, *8*, 686. [CrossRef]
7. Wang, L.; Wu, H.; Hu, Y.; Yu, Y.; Huang, K. Environmental Sustainability Assessment of Typical Cathode Materials of Lithium-Ion Battery Based on Three LCA Approaches. *Processes* **2019**, *7*, 83. [CrossRef]

*Review*

# A Comprehensive Review and Technical Guideline for Optimal Design and Operations of Fuel Cell-Based Cogeneration Systems

**Farah Ramadhani [1], Mohd Azlan Hussain [1,*] and Hazlie Mokhlis [2]**

1 Chemical Engineering Department, Faculty of Engineering, University of Malaya, Kuala Lumpur 50603, Malaysia; farahramadhani@siswa.um.edu.my
2 Electrical Engineering Department, Faculty of Engineering, University of Malaya, Kuala Lumpur 50603, Malaysia; hazli@um.edu.my
* Correspondence: mohd_azlan@um.edu.my

Received: 2 October 2019; Accepted: 10 December 2019; Published: 12 December 2019

**Abstract:** The need for energy is increasing from year to year and has to be fulfilled by developing innovations in energy generation systems. Cogeneration is one of the matured technologies in energy generation, which has been implemented since the last decade. Cogeneration is defined as energy generation unit that simultaneously produced electricity and heat from a single primary fuel source. Currently, the implementation of this system has been spread over the world for stationary and mobile power generation in residential, industrial and transportation uses. On the other hand, fuel cells as an emerging energy conversion device are potential prime movers for this cogeneration system due to its high heat production and flexibility in its fuel usage. Even though the fuel cell-based cogeneration system has been popularly implemented in research and commercialization sectors, the review regarding this technology is still limited. Focusing on the optimal design of the fuel cell-based cogeneration system, this study attempts to provide a comprehensive review, guideline and future prospects of this technology. With an up-to-date literature list, this review study becomes an important source for researchers who are interested in developing this system for future implementation.

**Keywords:** review; cogeneration; fuel cell; optimal design; guidelines

## 1. Introduction

The rapid increase of energy demand in conjunction with the depletion of oil and coal and the environmental threats to pollution over the world have led to an energy security issue. Researchers, scientists and engineers are making effort to find solutions by using more effective and efficient power generation systems or finding energy sources that are cleaner and renewable. The prospect in creating new technologies for energy generation purpose and utilizing cleaner energy sources have increased around the world by the commitment of countries to reduce their carbon emissions and to include the renewable energy sector into their energy plan [1,2].

In line with the development of energy generation systems, which are more efficient and reliable, the cogeneration system has played its role in power and heat production systems. The technology had been popular in 1977 using coal and oil as the fuels, but its prospect became more and more gloomy when the fuel price increased in 1980 [3]. However, this technology has gone back to be more popular in this last decade in line with the finding of new energy sources, which are renewable, cleaner and economically competitive. Currently, cogeneration systems can be derived not only using combustion engines or gas turbine but also employing fully renewable or semi-renewable energy sources such as photovoltaic thermal panels, Stirling engines and fuel cells.

Amongst the emerging technologies as the prime mover candidate for cogeneration systems, fuel cells are one of the most suitable devices that can generate electricity and heat continuously. Fuel cells act as an energy conversion device, which generates electricity from the thermodynamic and electrochemical reactions between hydrogen and oxygen. Along with the generated electricity from the fuel cells, they also generate heat, water and fewer carbon per kWh energy production compared to conventional combustion engines when using hydrocarbons as the fuel. The heat generated from the cell is potential to be used in the cogeneration system by producing hot water or converting it into cooling energy for room and water.

Based on its electrolyte technology and operating point, fuel cells have various types such as the proton exchange membrane fuel cell (PEMFC), direct methanol fuel cell (DMFC), alkaline fuel cell (AFC), phosphoric acid fuel cell (PAFC), molten carbonate fuel cell (MCFC), microbial fuel cell (MFC) and solid oxide fuel cell (SOFC) [4–6]. Amongst them, PEMFC being the low temperature fuel cell and SOFC as the high-temperature fuel cell are most popular to be employed as the prime mover in cogeneration systems. Application of these fuel cell types is not limited for residential use but also for industrial, public facilities and transportations [7].

Even though fuel cells are promising as a prime mover in cogeneration systems, the technology is expensive and has a long payback period, which is not economically competitive compared to other prime movers [8]. The research and development of new materials, which are cheaper and flexible with various fuels are needed to be done to reduce the investment cost of the fuel cells. Furthermore, the optimal design of the fuel cell-based cogeneration system has been proven to reduce the total cost and carbon emission generated by the system [9]. The optimal design of the fuel cell-based cogeneration system is also effective in tackling the size issue of the system capacity that leads to the energy-waste problem.

There has been a rise in the research, development and review of the fuel cell-based cogeneration system from year to year. Arsalis et al. [10] did a comprehensive review of fuel cell-based power and heat generation system which focused on the technology and configuration of the system. The study concerned two fuel cell types (PEMFC and SOFC) as the prime mover technology for the studied cogeneration system along with the thermal management for the system. Milcarek et al. [11] gave a review for the fuel cell-based cogeneration system covering the fundamental aspect on the future prospect of this system for commercialization. The study focused on the application of the cogeneration system for residential use only. Other reviews of the cogeneration systems not only focused on the fuel cell as the prime mover but also other technologies such as gas turbine, combustion engines, Stirling engine and renewable energy devices [3,12,13]. It can be concluded that reviews of fuel cell-based cogeneration systems are still limited. From our knowledge, there is no review that focused on the optimal design of fuel cell-based cogeneration system and guideline to design an optimal system based on its applications, energy requirements and various specific criteria.

Therefore, this study attempts to provide a comprehensive review of fuel cell-based cogeneration systems including its theoretical and working principle, research, development, commercialization, current state of the system and on the optimal design of the system. This study also provides guidelines for designing an optimal cogeneration system by using the fuel cell as the prime mover with its future prospects. An up-to-date summary of previous studies conducted in the past 5 years has also been included to give an insight for researchers who are interested in further studying the fuel cell-based cogeneration systems.

## 2. Overview of Fuel Cells and Cogeneration Systems

### 2.1. Fuel Cells: Working Principle and Types

All fuel cells have two porous electrodes called anode and cathode, which are separated by a dense electrolyte layer. They have similar characteristics to a battery in converting chemical primary sources into electrical energy through electrochemical reactions. The reactions occurring between hydrogen, oxygen and other oxidizing agents generate heat and water as the by-products and electricity as the primary product. In general, hydrogen as fuel moves through the porous anode while the oxygen as the oxidant transport through the porous cathode. In the interface between the anode and cathode, the hydrogen breaks up to $H^+$ ions and two electrons, which are absorbed to the electrode surface and pass through an external circuit to create direct current power as explained in the literature [11]. At the same time, the oxygen molecule at the porous anode combines with the two electrons from the electrode to form $O^{2-}$ ion, which diffuses to the electrolyte layer and reacts with $H^+$ ions to form water molecule.

The development of the electrolyte material enhances fuel cells to be fueled by other than pure hydrogen. Due to the high-cost of pure hydrogen, some fuel cells can be driven using hydrocarbon fuels. Hydrocarbons can be used via external reforming such as steam reforming or fuel combustion or via internal reforming on a catalyst layer with direct electro-oxidation [11]. Steam reforming is an endothermic reaction that reforms the hydrocarbon to hydrogen and syngas (CO). For several fuel cell types especially those that work at high temperature, the syngas can be used directly to form two electrons and carbon dioxide. Meanwhile, for low-temperature fuel cells, the gas must be processed into pure hydrogen through the water gas shift reaction where the syngas reacts to water to form pure hydrogen and carbon dioxide.

Fuel cells have also attracted much attention due to its environment friendly nature compared to the conventional generators, which generate harmful gases as by-products. According to Table 1, different types of fuel can be used to drive the fuel cells. Pure hydrogen is commonly used by low-temperature fuel cells such as alkaline fuel cell (AFC) and polymer electrolyte membrane fuel cell (PEMFC). The pure hydrogen itself can be produced from hydrocarbons, methanol or syngas. High-temperature types such as molten carbonate fuel cell (MCFC) and solid oxide fuel cell (SOFC) are more flexible in the use of the fuel. Furthermore, the fuel price can be competitive by using various types of hydrocarbon, biogas and natural gas.

Fuel cells can be categorized as pure renewable energy generation if pure hydrogen is used to drive the cells as they only produce water as the by-product [11]. However, the process of producing hydrogen, which mostly comes from the hydrocarbon reforming processes must be taken into consideration when calculating the life cycle assessment of the fuel cells. In several high-temperature fuel cells, the CO produced in the steam reforming process can be used directly and produces $CO_2$ as by-products along with water. However, compared to combustion engines, fuel cells are more environmentally friendly even though some small emissions of carbon and NOx may be produced during the reforming processes as much as having higher operating efficiency.

### *2.2. Cogeneration: System Components and Applications*

In several applications, especially for offices and residential homes, electricity is not the sole energy required. Other energies such as heating and cooling water are also needed continuously [14]. However, most office and residential buildings utilized the separated system (SP) in generating electricity, heating and cooling energies to meet those requirements, which caused inefficiency in energy usage and significantly raises the energy cost. Therefore, an integrated system that can cover more than one energy demand is desired to enhance the system efficiency, energy utilization and cost, using what is called the cogeneration system.

**Table 1.** Comparison between different type of fuel cell [15–18].

| Fuel Cell Type | PEMFC | AFC | DMFC | PAFC | MCFC | SOFC |
|---|---|---|---|---|---|---|
| Operating temp (°C) | 30–100 | 90–100 | 50–100 | 160–220 | 600–700 | 500–1000 |
| Electrical efficiency (%) | 30–40 | 60 | 20–25 | 40–42 | 43–47 | 50–60 |
| Energy conversion efficiency (heat and power) (%) | 85–90 | 85 | 85 | 85–90 | 85 | up to 90 |
| Typical stack size | <1–100 kW | 10–100 kW | Up to 1.5 kW | 50–1000 kW (250 kW module typical) | <1–1000 kW (250 kW module typical) | 5–3000 kW |
| Electrolyte | Solid polymeric membrane | Aqueous solution of potassium hydroxide soaked in a matrix | Solid organic polymer poly-perfluoro sulfonic acid | 100% phosphoric acid stabilized in an alumina-based matrix | Li2CO3/K2CO3 materials stabilized in an alumina-based matrix | Solid, stabilized zirconia ceramic matrix with free oxide ions |
| Fuels | Hydrocarbons or methanol | Pure hydrogen | Methanol | Hydrogen from natural gas | Natural gas, biogas, others | Natural gas or propane, hydrocarbons or methanol |
| Operational life cycle | 40,000–50,000 h (stationary) Up to 5000 h (mobile) | Up to 5000 h | 10,000–20,000 h | Up to 40,000 h | Up to 15,000 h | Up to 40,000 h |

Cogeneration system can be defined as the system that generates simultaneous power and heat from the same primary energy source [3]. The power generated includes mechanical, electrical or even fuel conversion chemically. On the other hand, the system also generates useful heat, which can be used for heating, cooling, distiller purposes or converted to electricity. Furthermore, cogeneration processes can produce three or more types of energy, which are called trigeneration and polygeneration system with additional components.

Cogeneration system consists of a single or hybrid energy source called the prime mover that generates one or two types of primary power simultaneously and consists of auxiliary components to recover the primary energy from the prime mover as depicted in Figure 1. In several applications, a cogeneration system is also equipped with storage devices such as hot water tank or battery. The storages are used to store excess energies generated by the system. By using this configuration, cogeneration can reach an efficiency of up to 80% compared to the single-power generation system [19].

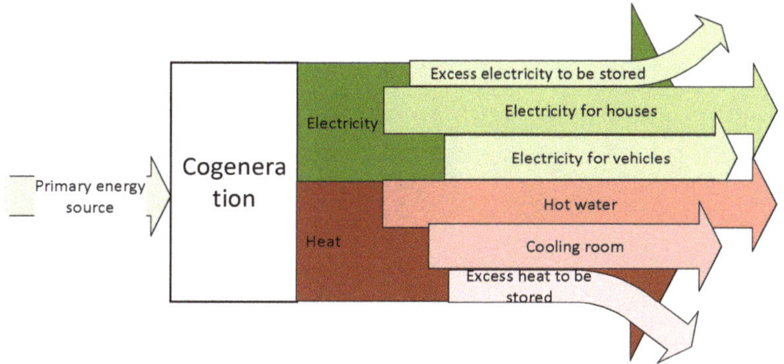


**Figure 1.** Cogeneration system layout.

Initially cogeneration system increased electricity generation by 58% in industrial plants [3] since the early century. However, due to economical, regulation and fuel availability issues, this system becomes less attractive for further development in the 1950s and accounted for only about 5% of the total electricity generation in the 1970s [3]. However, in the next few decades, implementation of cogeneration had been gaining attention again in line with the awareness of fuel depletion and environmental concern.

Combined heat and power (CHP) system is one of the most favorable types of cogeneration system, which generates electricity and heat. The CHP is efficient since it does not require additional fuels to produce heat power as in the separated system. The system was the first energy generation commercialized for residential applications, which had been successfully developed by several companies such as Hexis (Switzerland) and Ceres Power (UK), in partnership with British Gas and Ceramic Fuel Cells Ltd. (Australia) [20].

Currently, cogeneration systems have been designed and built for various other applications such as residential, industrial, public facilities and transportation. As the fuel cell is used as the prime mover, application for residential use as the stationary power system is more popular than others. In the industries, combinations of fuel cell fueled by biogas or syngas are also potential for waste-to-energy purposes in wastewater treatment (WWT) plant.

## 3. Current Developments of the Fuel Cell-Based Cogeneration Systems

The increased development of the fuel cell-based cogeneration system in the research and development sector as well as commercialization can be visualized by the rise of publications and commercial products in the last five years. Explanation of the current condition of the system development is discussed in these subsections below.

## 3.1. Research and Development Sector

Our review divides the research topics into three different types of fuel cell: polymer electrolyte membrane fuel cell (PEMFC), solid oxide fuel cell (SOFC) and other types of the fuel cell. The research and development of fuel cell-based cogenerations system as depicted in Figure 2 shows a positive trend in the past 10 years. It can be seen that both PEMFC and SOFC are the popular fuel cell implemented in cogeneration systems during that period.

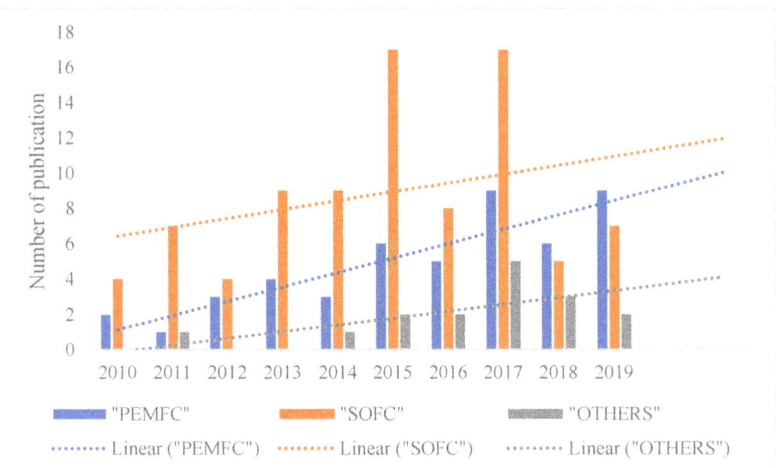


**Figure 2.** The research trends of the fuel cell-based cogeneration systems within the past 10 years.

Comparing these two, the applications involving PEMFC as the prime mover show a sharper increase as compared to the SOFC and others. One of the reasons is due to its flexibility of operation without any reforming and burning systems. The stability and load following capability of the PEMFC add more benefits to this type for small and mobile power generation. Moreover, further studies have developed the high-temperature proton membrane exchange fuel cell (HT-PEMFC), which can be more suitable for power and heat applications. The HT-PEMFC is seen to be popular and extensively developed since the past 5 years with 90% of system employment for CHP systems [21].

On the other hand, the increase of publication regarding SOFC-based cogeneration system is consistent from year to year. Not only developing the HT version of PEMC, but other studies also paid attention to the low-temperature solid oxide fuel cell (LT-SOFC). The LT-SOFC has been reported in several studies [22,23]. One of the reasons for decreasing the temperature is to reduce the material cost of the SOFC. The high temperature SOFC generates more heat and power but with increased cost in the electrolyte material as compared to the PEMFC. The high-temperature also causes the material to get cracked and degraded thus reducing the life cycle of the SOFC [24].

The other types of fuel cell such as PAFC, MCFC and DMFC have been reported in some studies [25–28]. The development of PAFC in Japan reported in [28] showed slow progress but promising for CHP systems in residential applications. However, not much attention has been given to further developments of other fuel cell types and this lack of study affects the progress of commercialization and their competitiveness in real applications.

## 3.2. Commercialization Sector

As a leader in this technology, Japan is pioneer in the development of fuel cells and cogeneration systems. As reported in the literature, the world's first residential proton exchange membrane fuel cell (PEMFC) CHP system in the Japanese market was built in 2009 [29]. It is planned that 5.3 million units of residential FC-CHP systems would be installed by 2030 to achieve Japan's Intended Nationally

Determined Contributions (INDC; a 26% reduction of total greenhouse gas (GHG) emissions by First Year (YF) 2030 compared with those in FY 2013) [30]. Furthermore, as Japan has succeeded to achieve GHG emission by 1270 $MtCO_2$/a in FY 2019, it has attained about 50% of the target of INDC [31].

In some of the European countries, the project H2home decentralized energy supply using hydrogen fuel cells is part of the HYPOS initiative (Hydrogen Power Storage and Solutions) [32]. In the building sector, proof of function has been provided in practice by the completed national project CALLUX (field test fuel cell for home ownership, 500 units in Germany) and the ongoing European project "Ene.Field" (which will deploy up to 1000 residential fuel cell micro-CHP installations across eleven key European countries). The European Commission set the greenhouse gas emissions and energy sustainability targets to be achieved by 2020: reducing by 20% the greenhouse gas emissions compared to 1990, reaching a share of 20% of renewable resources in the energy production and reducing by 20% the overall primary energy consumption compared to the projections made in 2007 [33].

Therefore, commercialization activities such as reducing the cost of the fuel cell system, increasing the electrical efficiency, increasing the energy efficiency in generating hydrogen, demonstrating the large-scale competitiveness of fuel cell and hydrogen technologies produced from primary renewable energy [34] will ensure that performance of the system fulfill the low-carbon economy target during this period up to 2050.

### 3.3. Governmental Support

In Japan, the promotion of SOFC micro CHP units involves an investment-based support scheme in the form of a capital grant. It reduces by half the initial cost of the generator, which is currently in use [35]. In Europe countries, a Feed-in Tariff scheme (price-based) was instead launched in 2010 in the United Kingdom (UK) where eligible generators are the micro-CHP units with a power output below 2 kW. The latter value has been chosen according to the cap given by the Feed-in Tariff actually adopted in the UK for 2 kW capacity for residential usage. Pellegrino et al. [35] studied the possible support by the UK governments in the fuel cell-based cogeneration system such as Capital grants, purchase and resale supports, Net metering support and two scenarios of feed-in-tariffs.

The United Nations Environmental Program has supported the Fuel cell installation with a total investment of $307.1 million in 2012, while the US Department of Energy (DOE) rolled out $9 million in grants to speed up the technology in June 2013 [34]. In China, the Ministry of Science and Technology of China, the Ministry of Finance of China, the Ministry of Industry and Information Technology of China and the National Development and Reform Commission of China have collaborated to develop new energy strategies by rolling out national grants focusing on fuel cells development and commercialization starting from 2012 [36]. Following this, other countries in Asia such as Malaysia has supported the utilization of renewable energy and development of hydrogen fuel cell through national grants given to universities [37] and feed-in-tariffs (FiT) scheme for residential applications [38,39].

## 4. Designing a Fuel Cell-Based Cogeneration System

Development of a better cogeneration system needs optimization of the overall system. Even though the cogeneration system is theoretically better than a separated system, the high-cost issue in the fuel cell development must be tackled by the proper design of the system. In optimizing the design, there are several steps to be followed as guidelines: modeling of the components, choosing the criteria for evaluation, evaluation of the system design, system control and management and optimization of the overall design. As depicted in Figure 3, these guidelines can be applied for any applications and system components to provide an optimal cogeneration system. The details of these steps will be explained in the subsections below.

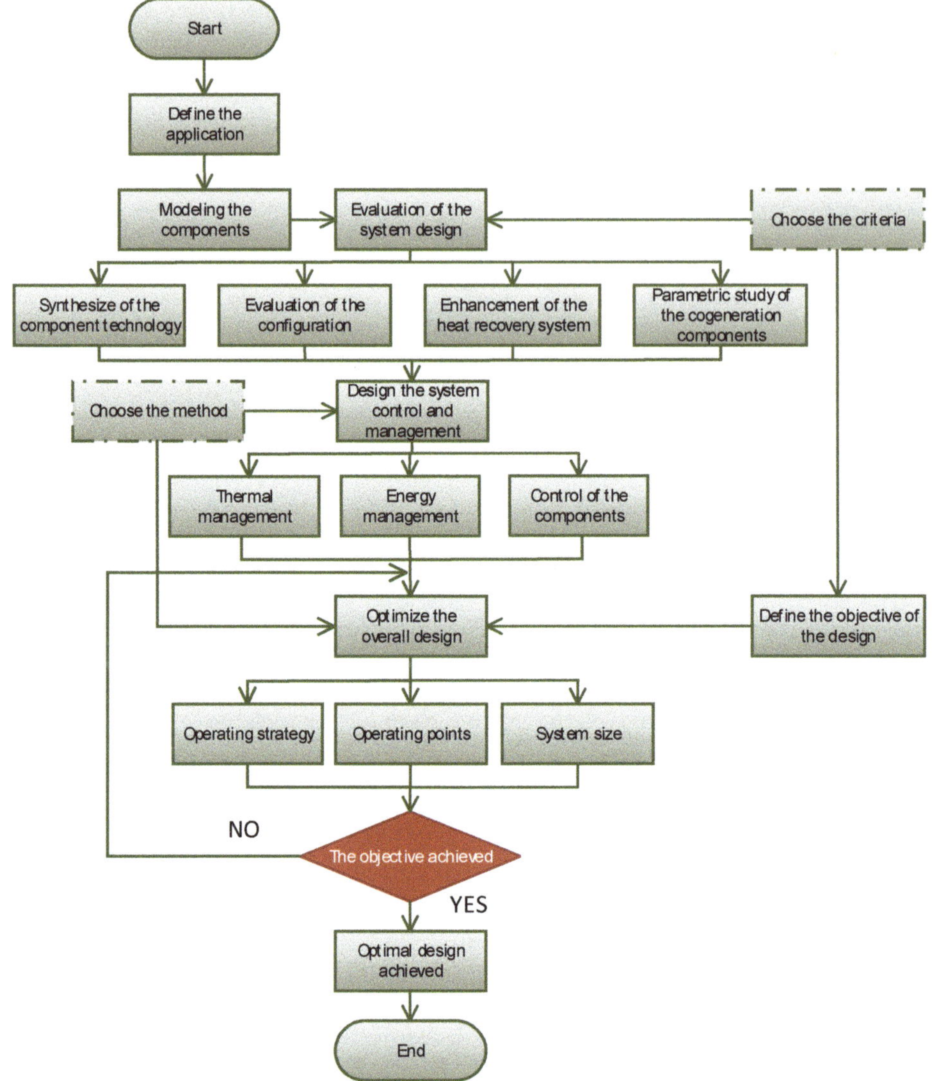


**Figure 3.** Guideline in designing an optimal cogeneration system with fuel cell as the prime mover.

### 4.1. Modeling the Components

In order to assess the performance of the cogeneration system, modeling the system components must be done first. The modeling part is always associated with the validation of the cogeneration component before going to be controlled and optimized. Most of the study focusing on the assessment of the cogeneration system built their system through mathematical modeling. With some assumptions and simplifications used, the model of components can validate the performance of the whole system. As a major prime mover of the cogeneration system, modeling of the fuel cell is vital to analyze the behavior of the component in generating power and heat for the cogeneration system.

In the modeling the cogeneration system, researchers have used several approaches such as 3D, 2D, 1D and even using 0D or black box predictions. Each approach is different in its complexity, accuracy and application. If the purpose is to achieve accuracy for detailed analysis, then the 3D approach is most suitable for the modeling approach. The mass, momentum and energy equations are presented in three dimensions with the heat transfer from the outer stack to surrounding surfaces [11]. This approach is mostly used for analyzing a single fuel cell where its geometry is appropriately discretized using the finite volume or finite difference method [11].

A 2D approach is simpler than the 3D approach because it neglects one dimension of fuel cell geometry and generates different models for different fuel cell geometries. However, the 1D approach is suitable for modeling of integrated fuel cells applications in a stack or combinations with other heating or cooling components in cogeneration systems. The 1D approach presents the fuel cell model in one direction for the variations of fuel cell temperature, pressure, concentration and other thermodynamic phenomena and material properties. In the literature, black box prediction models are also frequently used for analysis, control, management, evaluation and optimization studies in an integrated fuel cell-based system such as cogeneration.

Due to the complexity in integrating more than two components and applying adjusted operating strategies and sizes, the detailed fuel cell model must be simplified with a consequence in the reduction of accuracy. In order to enhance the accuracy of the model, most of the studies consider fuel cells as the prime mover in a cogeneration system and for more advanced systems, which include real experimental data from the literature to validate the cell model [39–42]. One example is that conducted by Asensio [43] that predicted the PEMFC system for optimal energy management using a black box model, and applied the adaptive neural network (ANN) combined with a 3D lookup table to predict the hydrogen consumed and output power of PEMFC in the cogeneration system.

### 4.2. Choosing the Criteria for Evaluation

While designing a cogeneration system, one, two or more criteria are used as the objective of the design. Based on the literature, criteria based on energy, economics and environmental are commonly used in the cogeneration design for evaluation of different configurations, parameter analysis, energy management and optimization of the system. In some studies, criteria used for the design is not limited to single criteria but also multi-criteria such as energy-environmental, exergoeconomic or eco-environmental parameters. The multi-criteria parameters are used to assess the system to deal with more than one criterion to be satisfied.

From the technical aspect, many studies in the literature commonly used efficiency of the system as the criterion [44,45]. Other studies use primary energy saving (PES) or primary energy consumption (PEC) as the energy criteria [25,46]. The values of the PES or PEC are obtained by subtracting the amount of primary energy or fuel used in the reference system (usually using a separated system) with the proposed cogeneration system. Besides using energy output and energy efficiency as the criteria, many studies in the literature focused on the second law of thermodynamics by using exergy as the criterion. Exergy is the available energy, which can be used from the consumption of primary energy including power and heat. The concept of exergy is more viable and practical in the cogeneration system since not only electric power is considered but also heating and cooling demands. Several studies in the literature also used exergy and efficiency as the criteria for the system performance achieved [45,47,48]. From the criteria, exergy destruction can also be used to indicate which part of the cogeneration components affects the system performance.

From the economical aspect, criteria such as energy cost, net present cost, payback period and total costs of the system are used in the cogeneration design. For the investment of the system and analysis of the system viability, the payback period is commonly used as the criterion. Some scenarios involving subsidiary from the government, tax reduction, net metering to feed-in-tariffs as incentives were studied to make the system more economically competitive compared to the conventional separated system (CSS) [35].

Environmental aspects have also been taken into account in achieving a cleaner environment and reducing the pollution caused by the energy sector. The older energy generation technologies using non-renewable sources have created negative impacts to the environment, thus the developments in the new emerging technologies are essential to achieve the sustainability of the energy. Reduction of carbon emission and other harmful gases are also other criteria, which are used in the cogeneration design. Furthermore, life cycle analysis (LCA) is also used to analyze the impact of the technology used in the cogeneration system to the environment, which is comprehensive since it covers all aspects from system production to system operation.

### 4.3. Evaluation of the System Design

Evaluations of the system consisting of system synthesis and assessment are important to analyze the actual performance of cogeneration systems. Evaluation and synthesis of the system include its configurations, prime mover types, heating/cooling devices, storages and other auxiliary components in the cogeneration system. Synthesis of the system defines the acceptability of each component in a cogeneration system in improving the performance of the system. System synthesis for cogeneration system has been made using the P-graph Fuzzy approach [49], TOPSIS [50] and MILP [51]. On the other hand, evaluation of the system can be performed by a parametric study to analyze the operating points of the components [27,40,45]. Furthermore, some modifications in the heat recovery system and system operation have also been done by [25,52]. Regarding the implementation of the cogeneration system, other studies compared the system for different climate conditions, places, and demand profiles [41,53,54].

Several works as reported in the literature, used a conventional separated system (CSS) and compared it with the proposed cogeneration [39,55,56]. The CSS uses different primary energy sources to provide electricity, cooling or heating for the users. Electric power is commonly provided from the national grid while heating or hot water is generated from a fuel driven boiler. Several studies found that the fuel cell-based cogeneration system is more promising to reduce primary energy consumption, energy cost and carbon emission generated by the system compared to the CSS [55,57,58].

### 4.4. System Control and Management

System operation in a cogeneration system plays an important role in reducing unused wasted energy that is generated by the prime mover and other components where, optimal energy management is able to reduce primary energy consumptions and operation costs. The operating strategies, which have been reported in the literature involve control and management for the prime mover, heating/cooling devices, storages and also dispatch mechanisms between cogeneration components in satisfying the load demands.

As the prime mover, a fuel cell can generate electricity and heat as long as the fuel is injected into the cell. However, the fluctuations in demand, fuel costs and other varying conditions affect the operation of the fuel cell. Therefore, fuel control is one of the options to optimize the utilization of the fuel depending on the power and heat required by the demands. Moreover, high-temperature fuel cells such as the SOFC and PAFC are very sensitive to the temperature shock caused by the high fluctuation of temperature during the operation. It needs thermal management to avoid material cracking and increase the life cycle of the cell.

Some energy management approaches related to energy storage control or demand control have also produced a better and efficient cogeneration system. The controls are also capable of reducing components capacity, thus reducing the investment and operational costs. The energy management approach in relation to storage control can also improve the reliability of the system, preventing system blackout and utilizing excess power generated from the supply side.

### 4.5. Optimization of the Overall Design

As the last step in designing the cogeneration systems, optimization approach can be conducted based on the fulfilled objectives. Optimization for the cogeneration system involves optimal operating strategy (OOS), optimal operating parameters (OOP) and optimal size of the cogeneration components

(OS). These three design objectives are significant variables in increasing the overall performance of the cogeneration system.

In designing the operating strategies, several system operations need an optimization approach to find the best strategy in their cogeneration designs. Scheduling and dispatching the energy from the cogeneration components to the demand side involve many combinations that have to be examined. In order to fulfill one or more objectives, a combined operating strategy can be the better choice. Therefore, the role of optimization in this case is to find the best operating strategy to be applied with the specific objectives as the requirement.

In terms of the operating parameters, optimization of those parameters can improve the performance of the cogeneration components in reducing the fuel usage, decreasing its costs or generating less carbon emission. As the prime mover, fuel cell parameters such as temperature, hydrogen flow, steam to carbon ratio and pressure can be optimized to improve the flexibility in its operation and reducing its primary energy consumption. Meanwhile, other component parameters for generating hot water, cooling or hydrogen can also be used to reduce the costs of the components and increase the value of the cogeneration system.

Several studies in the literature have also focused on optimizing the size of the various components of the cogeneration system [38,59,60]. From these studies, comparisons of various configurations on the cogeneration performance is essential to avoid oversizing or under sizing of the system for the specific energy demand and applications.

## 5. Summary of the Gathered Literature in the Past 5 Years

As presented in Table 2, the publication summary shows an intense increase in research and development for fuel cell-based cogeneration system in the last 5 years. There are several important summaries to be extracted from the Table. Firstly, PEMFCs and SOFCs are still the popular fuel cells for cogeneration systems and will be further developed for use in cogeneration systems. In the future, steady-state and linear models are mostly used for the modeling process of the system. On the other hand, studies that concern in the model prediction are limited although the predicted models have some advantages in simplifying the mathematical equations used in the modeling process and closer to the real performance when using real experimental data as the reference. A few numbers of study that focused on the control and energy management strategy was reported from the literature.

Furthermore, the topics that studied the hybrid configurations between fuel cell and renewable energy devices are also limited. Most of the studies found in the literature used the fuel cell as the sole prime mover in the cogeneration system. Only several studies combined between two types of fuel cell [25,26,28,57,61,62] and the combinations between the fuel cell and other energy conversion devices such as photovoltaic, electrolyzer, thermoelectric and batteries as the storage were reported in [46,63–65].

In terms of its applications and designed scenario, most of the cogeneration systems were implemented for residential and building sectors while the use in the transportation sector and mobile power generation have not been found. Furthermore, the number of studies that implemented the cogeneration system for providing other than power and heat (example treated water, cooling, hydrogen, oxygen, etc.) is still limited. It can also be seen that very few of the studies consider the external support from the society or impacts of the cogeneration on the society as the feasibility study and only a few studies included government support as the assessment scenario [19,38] in the design of the cogeneration system.

**Table 2.** Summary of studies in fuel cells-based cogeneration system design in the past 5 years (2015–2019).

| Authors | Prime Mover | Application | Designing Parts | | | | Criteria Used |
|---|---|---|---|---|---|---|---|
| | | | Modeling | Control/Management | Evaluation of System | Optimization | |
| Wakui et al. [25] | SOFC, PEFC | Residential | NL-Quasi steady state | RTC using MILP | 4 cases of operation | - | PEC (TCN) |
| Sarabchi et al. [66] | PEMFC (HT) | - | Steady state, EES | - | √ | MO | Energy and exergy efficiencies (TCN), specific costing (ECO), $MSE_{CO_2}$ (ENV) |
| Nalbant et al. [45] | PEMFC (HT) | - | Steady state, EES | - | Parameter analysis | - | Energy and exergy efficiencies (TCN), |
| Luo and Fong [52] | SOFC | - | 2D dynamic | | Configurations of bottoming cycle | - | Efficiency, PI, CR (TCN) |
| Löbberding and Madlener [67] | PEMFC | Residential | Steady state simulation | EO and HO operations | Economic analysis | - | NPV(ECO) and competitiveness |
| Kwan et al. [46] | PEMFC-TED | - | Steady state simulation | EMS with mode decision | - | - | PEC efficiency (TCN) |
| Jung et al. [53] | SOFC | Residential | TRNSYS | Thermal load following | Weather cons. and surplus elect. | - | PEC (TCN), OC & PP (ECO), $CO_2E$ (ENV) |
| Jin et al. [68] | PEMFC | - | Quantitative model | - | Configurations of extraction | - | COP (TCN) |
| Huang et al. [69] | FC | Residential | MINLP | PLR, Scheduling | - | PSO-SQP for scheduling | COE (ECO) |
| Guo et al. [40] | PEMFC (HT) | - | 1D isothermal model | - | Parametric studies | - | DL, Humidity, Heat loss, heat coefficient, |
| Marcoberardino et al. [41] | PEMFC | Residential | Aspen plus | - | Different configurations and scenarios | - | ES (ECO) |
| Cinti et al. [44] | SOFC | - | Cycle tempo | - | Effect of hythane as fuel | - | Efficiency, EI (TCN) |

**Table 2.** *Cont.*

| Authors | Prime Mover | Application | Designing Parts | | | | Criteria Used |
|---|---|---|---|---|---|---|---|
| | | | Modeling | Control/Management | Evaluation of System | Optimization | |
| Bachmann et al. [70] | PEMFC and SOFC | Residential | Environmental model | - | 4 Technologies of CHP | - | LCA, LCIA (ENV) |
| Roshandel et al. [71] | SOFC | Residential | Steady state | √ | 4 hybrid systems | MO | LCOE (ECO), $CO_2E$ (ENV) |
| Romdhane et al. [72] | PEMFC | Residential | Steady state using MATLAB | Heat-led Electric-led | √ | - | EFF-PES- (TCN), $CO_2E$ (ENV) |
| Facci et al. [54] | PEMFC | Buildings | Dynamic, AspenPlus | Cost and PEC minimizations | 5 buildings, 5 climate conditions | - | PEC (TCN), COE-PBP (ECO) |
| Yoda et al. [73] | SOFC | Residential | - | - | Real system of CHP | - | EFF-DRB (TCN), Cost reduction (ECO) |
| Wakui et al. [25] | SOFC and PEFC | Residential | MILP | Scheduling using MILP, operation control under receding horizon | 4 types of fuel cell and 4 storages | - | PEC (TCN), DRN |
| Baldi et al. [59] | SOFC and PEMFC | Residential | Linear model | - | - | Sizing using MILP | Efficiency (TCN), Investment cost (ECO) |
| Wu et al. [27] | PAFC-TEG | - | Steady state model | - | Parametric study of T, P, m, K, c1 and c2 | - | Current and power densities, efficiency (TCN) |
| Spazzafumo [74] | MCFC and SOFC | - | AspenOne | - | Evaluation of pressure composition | - | Efficiency (TCN) |
| Perna et al. [75] | SOFC | - | Numerical model using AspenPlus | - | Evaluation of GT pressure and S/C | - | Power and efficiency (TCN) |
| Ozawa and Kudoh [30] | PEMFC and SOFC | Residential | Linear model | OOP with MILP | Evaluation based on energy demand types | - | NPC (ECO), GHG emission (ENV) |

**Table 2.** *Cont.*

| Authors | Prime Mover | Application | Designing Parts | | | | Criteria Used |
|---|---|---|---|---|---|---|---|
| | | | Modeling | Control/Management | Evaluation of System | Optimization | |
| Herrmann et al. [32] | PEMFC | Residential | TRNSYS | - | Evaluation of CHP configurations | - | Primary energy, efficiency, usable energy (TCN), total cost (ECO), $CO_2$ Emissions |
| Mamaghani et al. [60] | PEMFC (HT) | - | Steady state model | - | - | MO using Pareto frontier GA | Thermal power and net power output, net electrical and thermal efficiency (TCN) |
| Giarolla et al. [76] | SOFC | Industrial (WWT) | MILP model | - | Evaluation of 3 scenarios and 5 configurations of the system | OOP using GAMS and CPLEX solver | LCOE (ECO) |
| Marcoberardino et al. [77] | PEMFC | Residential | Energy balance | - | Evaluation of system configuration | - | Target cost (ECO) LCA (ENV) |
| Chitsaz et al. [78] | SOFC | - | Steady state using MATLAB | - | Evaluation between 2 configurations | Optimal value of design parameters (MO) | Exergy efficiency (TCN), $CO_2$ gas emission (ENV) |
| Budak and Devrim [79] | PEMFC (LT and HT) | - | Experimental study | - | Performance test of two PEMFC types | - | SPT (ECO) |
| Asensio et al. [43] | PEMFC | - | Prediction model using ANN and 3D lookup table | - | - | - | Hydrogen flowrate and efficiency (TCN) |
| Aki et al. [28] | PEFC | Residential | MILP model | EMS prediction and OOP | - | - | Energy cost (ECO), PEC (TCN), $CO_2E$ (ENV) |

**Table 2.** *Cont.*

| Authors | Prime Mover | Application | Designing Parts | | | | Criteria Used |
|---|---|---|---|---|---|---|---|
| | | | Modeling | Control/Management | Evaluation of System | Optimization | |
| Wakui et al. [25] | PEFC and SOFC | Residential | MILP model | Energy demand prediction using SVR | - | OOP scheduling | Energy cost (ECO), PEC (TCN), $CO_2E$ (ENV) |
| Vialetto et al. [58] | SOFC | Residential and transportation | Linear model | Three operating strategies | Comparison between 3 systems | - | PES (TCN), EAC (ECO) |
| Tanaka et al. [80] | SOFC | Academic institution | Linear model | On-off control | Evaluation of system configuration | - | PES (TCN) |
| Mehr et al. [81] | SOFC | Industrial (WWT) | Steady state model | - | Evaluation of system configuration | - | Efficiency, ANGR (TCN) LCOE, PBT, NPV (ECO) |
| Kupecki et al. [82] | SOFC | - | Aspen HYSYS and Experimental data | - | Parametric study | - | Efficiency, output power (TCN) |
| Karami et al. [83] | PEMFC | Residential | Energy balance model | EMS with scheduling | - | OOP using CCA | OC (ECO) |
| Hosseinpour et al. [48] | SOFC | - | EES | - | Parametric study (j, Ti, CR, εr) | OOP using DSM | Energy and exergy efficiency (TCN) |
| Mejia et al. [84] | SOFC | Residential | Experimental data | - | Evaluation of 3 configuration cases | - | Grid imported, peak power (TCN), Total cost (ECO), GHG emission (ENV) |
| Hajabdollahi et al. [85] | SOFC | Residential | Energy balance | - | - | OOP using MOGA | Exergy efficiency (TCN), TCR (ECO) |
| Mamaghani et al. [86] | PEMFC (HT) | - | 1D steady state model | - | - | OOP using MOGA | Electrical efficiency, thermal generation, electrical power generation |

**Table 2.** *Cont.*

| Authors | Prime Mover | Application | Designing Parts | | | | Criteria Used |
|---|---|---|---|---|---|---|---|
| | | | Modeling | Control/Management | Evaluation of System | Optimization | |
| Eveloy et al. [87] | SOFC | Industrial (Desalination plant) | Aspen plus and FORTRAN | - | Evaluation of system | OOP using GA and TOPSIS | Exergy efficiency (TCN), TCR (ECO) |
| Anyenya et al. [88] | SOFC | Industrial (in situ oil shale) | Aspen plus | - | Parametric study | - | Electric power, stack temperature, HE ratio, efficiency (TCN) |
| Zhang et al. [89] | SOFC | - | Steady state | - | Parametric study | - | Efficiency, power density (TCN) |
| Reyhani et al. [90] | SOFC | Industrial (MED) | Steady state using MATLAB | - | Evaluation of configuration | MO using GA | Exergy efficiency (TCN), ACS (ECO) |
| Pohl et al. [91] | PEMFC (HT) | Residential | Linear model | Heat-driven, on-off switch | Evaluation of system performance | - | PES, Degree of coverage (TCN) |
| Misra et al. [92] | FC | Residential | Steady state using HOMER | - | Evaluation of the system performance | - | NPC (ECO) |
| Khani et al. [42] | SOFC | - | Steady state model using EES | - | - | OOP in MO using GA MATLAB | Exergy efficiency (TCN), SUCP (ECO) |
| Khani et al. [93] | SOFC | - | Steady state model using EES | - | Evaluation of the configurations | OOP in MO using GA MATLAB | Exergy efficiency, exergy destruction (TCN), SUCP (ECO) |
| Isa et al. [38] | PEMFC | Hospital | Steady state using HOMER | - | Evaluation of the configurations | Size optimization | Energy allocation (TCN) LCOE, TNPC, LCC, and salvage cost (ECO), $CO_2E$ (ENV) |
| Hassanzadeh et al. [55] | SOFC | Residential | Energy and exergy balance | - | Evaluation of the system configurations | MO | Power and production, exergy destruction (TCN) |

**Table 2.** *Cont.*

| Authors | Prime Mover | Application | Designing Parts | | | | Criteria Used |
|---|---|---|---|---|---|---|---|
| | | | Modeling | Control/Management | Evaluation of System | Optimization | |
| Mamaghani et al. [21] | PEMFC (HT) | - | 1D steady state model | - | Evaluation of the system conditions | OOP in MO using GA | Net electrical efficiency (TCN), TCC (ECO) |
| Fong and Lee [94] | SOFC | Residential | Dynamic, TRNSYS | - | Parametric study | - | PEC (TCN), PP (ECO), $CO_2E$ (ENV) |
| Assaf and Shabani [95] | PEMFC | Residential | Dynamic, TRNSYS | - | Evaluation of the system configurations | Size optimization using DSM | Power and heat generation (TCN), NPC (ECO) |
| Windeknecht and Tzscheutschler [96] | SOFC | Residential | SimulationX | - | Evaluation of the system configurations | - | PEC (TCN), ATC (ECO) |
| Vialetto and Rokni [97] | SOFC | Residential | DNA | Electric equivalent load (EEL) | Evaluation of the system configurations | - | PES (TCN), NPV, PP (ECO) |
| Ullah et al. [98] | SOFC | - | Experimental data | - | Parametric study | - | Output power and efficiency (TCN) |
| Shariatzadeh et al. [99] | SOFC-SOEC | Solar chimney | Steady state | - | - | OOP and size using GA (SO) | Total cost (ECO) |
| Pellegrino et al. [35] | SOFC | Residential | Linear model | Continuous, modulations, controlled output | Evaluation of the system conditions | Size optimization using direct search | PP, total retail cost (ECO) |
| Napoli et al. [61] | SOFC and PEMFC | Residential | Linear model | Modulation strategies | Evaluation of the system conditions | - | PES (TCN), NPV (ECO) |
| Liso et al. [100] | SOFC | Single house | Steady state | - | Evaluation based on fuel types | Tank sizing | Heat recovery (TCN) |
| Kupecki et al. [101] | SOFC | - | Steady state and experimental models | - | Parametric study | - | Electrical efficiency (TCN) |

**Table 2.** *Cont.*

| Authors | Prime Mover | Application | Designing Parts | | | | Criteria Used |
|---|---|---|---|---|---|---|---|
| | | | Modeling | Control/Management | Evaluation of System | Optimization | |
| Ham et al. [102] | PEMFC | Residential | Steady state and empirical data (black box) | - | Evaluation of system performance | - | Net output power, Net heat power (TCN) |
| Elmer et al. [103] | SOFC | Single home | Real experimental data | - | Eco-environmental assessment | - | $CO_2E$ (ENV), Cost reduction (ECO) |
| Cappa et al. [104] | PEMFC | Residential | Steady state | Thermal tracking | Evaluation of the system configurations | OOC using dynamic programming | PEC (TCN), NPV (ECO) |
| Canelli et al. [105] | PEMFC | House and office | Dynamic, TRNSYS | Load sharing | Evaluation of system configurations | - | PES (TCN), Operational cost reduction (ECO), $CO_2eq$ (ENV) |
| Borji et al. [106] | SOFC | - | 1D model | - | Parametric study | OOP using NSGA | Output power and efficiency (TCN) |
| Arsalis et al. [107] | PEMFC | Residential | 1D model | - | Evaluation of system configurations | OOP and size using EES | System efficiency (TCN) LCC (ECO) |
| Antonucci et al. [108] | SOFC | Residential | Steady state and experimental data | Thermal standby and electric standby | Evaluation of system configurations | - | PES (TCN), Specific costs (ECO) |
| Akikur et al. [39] | RSOFC | Single house | Steady state | PV-SOSE, SOFC and PV-SOFC modes | Evaluation of system configurations | Sizing using DSM | Efficiency (TCN) COE, PP (ECO), $CO_2E$ (ENV) |

## 6. Future Directions of Fuel Cells Application in Energy Generation Systems

Based on the current status of fuel cell developments in cogeneration systems and the review done, several promising directions for future developments of the system can be obtained. In terms of the research topic regarding the optimum system, the study that involves monitoring and predictions aspects are potential in the design of optimal cogeneration system. The monitoring and predictions are not only conducted for the cogeneration components, but also for the demand profiles and the operation of the system. These topics could increase the value of the optimum system since the data collected is not based on the assumptions but real experimental data. Predicted system components and the demands also simplify the analysis of the system performance and lessen the complexity in the interactions between the cogeneration components since no mathematical model is used.

In line with the monitoring and prediction of the system, experimental studies involving the real fuel cell-based cogeneration system is valuable to analyze the durability of the system. A couple of studies that have implemented the cogeneration system for a real implementation can be used as references [109,110]. Since the PEMFC and SOFC are well known as the prime mover in cogeneration systems, the finding of its commercialized products is easier to obtain. However, finding the commercialized products for the other types of fuel cell is challenging, thus experimental studies regarding these other types of fuel cell is highly promising.

Moreover, the cost issue regarding fuel cell development can also be tackled by finding technologies for fuel reforming and using various types of fuel. Some studies have started to develop syngas and various hydrocarbons as fuels for driving the cogeneration system [75,90]. Other studies focused on the new materials of the electrolyte of the fuel cells to reduce the investment cost and increase the life cycle [111,112].

Besides using various types of fuel, the cogeneration performance can also be increased by finding technologies in optimizing electricity production from the fuel cells. Combination between fuel cells and other power generators as a hybrid prime mover is the key to doubling the electric power generation and reducing the size of the fuel cell. In terms of hybrid the cogeneration system with other energy conversion devices, several promising units such as solar rechargeable, thermoelectric, electrolyzer with solid oxide electrolysis cell and flow batteries can be coupled with the system to increase the fuel utilization and system capacity with valuable costs [27,63,113].

For system operations, energy management strategy combined with optimal operation parameters and system predictions seems important to be developed, which have shown good results in reducing primary energy consumption as well as its operating cost and carbon emission from the cogeneration system. The predictions in the demand can also tackle the energy loss issue and increase the reliability of the system.

Lastly, applications for waste-to-energy usage have huge potential for the further development of the fuel cell-based cogeneration system where these newly innovative systems can be economically competitive in the commercial and government sector.

## 7. Conclusions

Based on the scientific indicator that was presented in the research review, PEMFC and SOFC were the two well-known and most applied fuel cells among others. Current developments of those fuel cells show that they were more being widely used especially with further improvement of its material to increase its durability with higher temperature ranges. Furthermore, being one of the focuses of this study, a guideline to develop an optimal fuel cell cogeneration system was also presented. The guidelines start from the modeling of the components, assessment of the system design, designing the operating strategy and optimization of the overall design involving operating parameters and size of the components. Through the guidelines given, optimal design can be done comprehensively using different specific applications and criteria for the implementation of the cogeneration system. Numerous publications for the last five years can be a good point of reference to design an optimal cogeneration system with various approaches, objectives and applications. Those publications also indicated the various ways to increase system performance, reduce system cost and emissions of the systems and give more insight for the researchers and developers who are interested to work in this area in the future.

For power generation, fuel cell-based cogeneration system has a better future compared to the conventional heat engine-based technology. From this study it also can be seen that various hydrocarbon fuels have been utilized to replace the utilization of pure hydrogen as to reduce the fuel cost with various materials chosen to increase the temperature range and durability of the fuel cells. Application of cogeneration system can be explored widely not only for stationary but also for mobile power generation uses in the future.

**Author Contributions:** All authors have contributed to this research article. M.A.H. has contributed in conceptualization and editing, while H.M. has revised technical terms in this article. As the main author, F.R. has contributed in conceptualization, writing and editing processes.

**Funding:** This research received no external funding.

**Conflicts of Interest:** The authors declare no conflict of interest.

## Abbreviations/Symbols

| | |
|---|---|
| ACS | Annualized cost system |
| AFC | Alkaline fuel cell |
| ANGR | Annual natural gas reduction |
| ANN | Adaptive neural network |
| CCA | Colonial competitive algorithm |
| CHP | Combined heat and power |
| CO2E | Carbon dioxide emission |
| COP | Coefficient of performance |
| CR | Contribution ratio |
| CSS | Conventional separated system |
| DL | Doping level |
| DMFC | Direct methanol fuel cell |
| DRB | Durability |
| DSM | Direct search method |
| EAC | Estimated annual cost |
| ECO | Economic aspect |
| EES | Engineering equation solver |
| EFF | Efficiency |
| EI | Energy index |
| EMS | Energy management strategy |
| ENV | Environmental aspect |
| EO | Electric operation |
| ES | Energy saving |
| GHG | Greenhouse gas |
| HE | Heat-to-electric |
| HO | Heat operation |
| HT | High temperature |
| LCA | Life cycle analysis |
| LCIA | Life cycle integrated analysis |
| LCOE | Levelized cost of energy |
| LT | Low temperature |
| MCFC | Molten carbonate fuel cell |
| MED | Multi-effect distillation |

| | |
|---|---|
| MFC | Micro-bacterial fuel cell |
| MILP | Mixed integer linear programming |
| MINLP | Mixed integer non-linear programming |
| MO | Multi-objective |
| MOGA | Multi-objective genetic algorithm |
| MSE | Mass specific emission |
| NL | Nonlinear |
| NPC | Net present cost |
| NPV | Net present cost |
| OC | Operational cost |
| OOP | Optimal operating parameter |
| PAFC | Phosphoric acid fuel cell |
| PBT | Payback time |
| PEC | Primary energy consumption |
| PEMFC | Proton exchange membrane fuel cell |
| PES | Primary energy saving |
| PI | Performance index |
| PLR | Part load ratio |
| PP | Payback period |
| RTC | Real-time control |
| S/C | Steam to carbon ratio |
| SOFC | Solid oxide fuel cell |
| SP | Separated system |
| SPT | System payback time |
| SUCP | Sum of unit cost of products |
| SVR | Support vector regression |
| TCN | Technical aspect |
| TCR | Total cost rate |
| TED | Thermoelectric device |
| WWT | Wastewater treatment |
| $T$ | Temperature |
| $P$ | Pressure |
| $m$ | Thermoelectric elements |
| $K$ | Conductance |
| $c1$ | Regenerative losses |
| $c2$ | Heat-leakage losses |
| $j$ | Current density |
| $Ti$ | Inlet temperature of SOFC |
| $\varepsilon r$ | Regenerator effectiveness |

## References


1. Busu, M. The Role of Renewables in a Low-Carbon Society: Evidence from a Multivariate Panel Data Analysis at the EU Level. *Sustainability* **2019**, *11*, 5260. [CrossRef]
2. Zhao, W.; Zou, R.; Yuan, G.; Wang, H.; Tan, Z. Long-Term Cointegration Relationship between China's Wind Power Development and Carbon Emissions. *Sustainability* **2019**, *11*, 4625. [CrossRef]
3. Frangopoulos, C.A. Cogeneration Technologies, Optimisation and Implementation. In *Iet Energy Engineering*; The Institution of Engineering and Technology: London, UK, 2017; Volume 87.
4. Weng, G.-M.; Li, C.-Y.V.; Chan, K.-Y. Hydrogen battery using neutralization energy. *Nano Energy* **2018**, *53*, 240–244. [CrossRef]
5. Weng, G.-M.; Li, C.-Y.V.; Chan, K.-Y. An Acid–Base Battery with Oxygen Electrodes: A Laboratory Demonstration of Electrochemical Power Sources. *J. Chem. Educ.* **2019**, *96*, 1701–1706. [CrossRef]
6. Maffei, N.; Pelletier, L.; Charland, J.; McFarlan, A. An intermediate temperature direct ammonia fuel cell using a proton conducting electrolyte. *J. Power Sources* **2005**, *140*, 264–267. [CrossRef]
7. Kordesch, K.; Simader, G. *Fuel Cells and Their Applications*; Wiley-VCH: Weinheim, Germany, 1996.

8. Ramadhani, F.; Hussain, M.A.; Mokhlis, H.; Fazly, M.; Ali, J.M. Evaluation of solid oxide fuel cell based polygeneration system in residential areas integrating with electric charging and hydrogen fueling stations for vehicles. *Appl. Energy* **2019**, *238*, 1373–1388. [CrossRef]
9. Al Moussawi, H.; Fardoun, F.; Louahlia, H. Selection based on differences between cogeneration and trigeneration in various prime mover technologies. *Renew. Sustain. Energy Rev.* **2017**, *74*, 491–511. [CrossRef]
10. Arsalis, A. A comprehensive review of fuel cell-based micro-combined-heat-and-power systems. *Renew. Sustain. Energy Rev.* **2019**, *105*, 391–414. [CrossRef]
11. Milcarek, R.J.; Ahn, J.; Zhang, J. Review and analysis of fuel cell-based, micro-cogeneration for residential applications: Current state and future opportunities. *Sci. Technol. Built Environ.* **2017**, 1–20. [CrossRef]
12. Isa, N.M.; Tan, C.W.; Yatim, A.H.M. A comprehensive review of cogeneration system in a microgrid: A perspective from architecture and operating system. *Renew. Sustain. Energy Rev.* **2018**, *81*, 2236–2263. [CrossRef]
13. Murugan, S.; Horák, B. A review of micro combined heat and power systems for residential applications. *Renew. Sustain. Energy Rev.* **2016**, *64*, 144–162. [CrossRef]
14. Ahmadi, P. *Modeling, Analysis and Optimization of Integrated Energy Systems for Multigeneration Purposes*; University of Ontario Institute of Technology: Oshawa, ON, Canada, 2013.
15. Elmer, T. *A Novel SOFC Tri-Generation System for Building Applications*; University of Nottingham: Nottingham, UK, 2015.
16. Ellamla, H.R.; Staffell, I.; Bujlo, P.; Pollet, B.G.; Pasupathi, S. Current status of fuel cell based combined heat and power systems for residential sector. *J. Power Sources* **2015**, *293*, 312–328. [CrossRef]
17. US Department of Education. *(EERE), U.D.E.E.a.R.E., Ed.*; Energy Efficiency and Renewable Energy: Washington, DC, USA, 2007.
18. Kimiaie, N.; Wedlich, K.; Hehemann, M.; Lambertz, R.; Müller, M.; Korte, C.; Stolten, D. Results of a 20,000 h lifetime test of a 7 kW direct methanol fuel cell (DMFC) hybrid system–degradation of the DMFC stack and the energy storage. *Energy Environ. Sci.* **2014**, *7*, 3013–3025. [CrossRef]
19. Chen, J.M.P.; Ni, M. Economic analysis of a solid oxide fuel cell cogeneration/trigeneration system for hotels in Hong Kong. *Energy Build.* **2014**, *75*, 160–169. [CrossRef]
20. Irvine, J.T.S.; Connor, P. *Solid Oxide Fuels Cells: Facts and Figures: Past Present and Future Perspectives for SOFC Technologies*; Springer: London, UK, 2012.
21. Haghighat Mamaghani, A.; Najafi, B.; Casalegno, A.; Rinaldi, F. Long-term economic analysis and optimization of an HT-PEM fuel cell based micro combined heat and power plant. *Appl. Therm. Eng.* **2016**, *99*, 1201–1211. [CrossRef]
22. Mosaffa, A.H.; Farshi, L.G. Thermodynamic and economic assessments of a novel CCHP cycle utilizing low-temperature heat sources for domestic applications. *Renew. Energy* **2018**, *120*, 134–150. [CrossRef]
23. Abdullah, T.; Liu, L. Simulation-based microstructural optimization of solid oxide fuel cell for low temperature operation. *Int. J. Hydrog. Energy* **2016**, *41*, 13632–13643. [CrossRef]
24. Ramadhani, F.; Hussain, M.A.; Mokhlis, H.; Hajimolana, S. Optimization strategies for Solid Oxide Fuel Cell (SOFC) application: A literature survey. *Renew. Sustain. Energy Rev.* **2017**, *76*, 460–484. [CrossRef]
25. Wakui, T.; Sawada, K.; Kawayoshi, H.; Yokoyama, R.; Iitaka, H.; Aki, H. Optimal operations management of residential energy supply networks with power and heat interchanges. *Energy Build.* **2017**, *151*, 167–186. [CrossRef]
26. Wakui, T.; Kawayoshi, H.; Yokoyama, R. Optimal structural design of residential power and heat supply devices in consideration of operational and capital recovery constraints. *Appl. Energy* **2016**, *163*, 118–133. [CrossRef]
27. Wu, M.; Zhang, H.; Zhao, J.; Wang, F.; Yuan, J. Performance analyzes of an integrated phosphoric acid fuel cell and thermoelectric device system for power and cooling cogeneration. *Int. J. Refrig.* **2018**, *89*, 61–69. [CrossRef]
28. Aki, H.; Wakui, T.; Yokoyama, R.; Sawada, K. Optimal management of multiple heat sources in a residential area by an energy management system. *Energy* **2018**, *153*, 1048–1060. [CrossRef]

29. Partners, E.-F. Accumulating Numbers of Residential Fuel Cell Ene-Farm Exceeded 200,000 Units. (In Japanese)
30. Ozawa, A.; Kudoh, Y. Performance of residential fuel-cell-combined heat and power systems for various household types in Japan. *Int. J. Hydrog. Energy* **2018**, *43*, 15412–15422. [CrossRef]
31. Climate Governance in Japan. Available online: https://climateactiontracker.org/countries/japan/ (accessed on 3 November 2019).
32. Herrmann, A.; Mädlow, A.; Krause, H. Key performance indicators evaluation of a domestic hydrogen fuel cell CHP. *Int. J. Hydrog. Energy* **2018**, *44*, 19061–19066. [CrossRef]
33. Commission, E. Europe 2020–EU-wide Headline Targets for Economic Growth. (n.d.) Available online: http://ec.europa.eu/europe2020/europe-2020-in-a-nutshell/targets/index_en.htm (accessed on 23 June 2019).
34. Niakolas, D.K.; Daletou, M.; Neophytides, S.G.; Vayenas, C.G. Fuel cells are a commercially viable alternative for the production of "clean" energy. *Ambio* **2016**, *45*, S32–S37. [CrossRef]
35. Pellegrino, S.; Lanzini, A.; Leone, P. Techno-economic and policy requirements for the market-entry of the fuel cell micro-CHP system in the residential sector. *Appl. Energy* **2015**, *143*, 370–382. [CrossRef]
36. Lu, Y.; Cai, Y.; Souamy, L.; Song, X.; Zhang, L.; Wang, J. Solid oxide fuel cell technology for sustainable development in China: An over-view. *Int. J. Hydrog. Energy* **2018**, *43*, 12870–12891. [CrossRef]
37. Mohamed, W.A.N.W.; Atan, R.; Sin, Y.T. Current and possible future applications of hydrogen fuel cells in Malaysia. In Proceedings of the International Conference on Advances in Mechanical Engineering (ICAME), Kuala Lumpur, Malaysia, 24–25 June 2009.
38. Isa, N.M.; Das, H.S.; Tan, C.W.; Yatim, A.H.M.; Lau, K.Y. A techno-economic assessment of a combined heat and power photovoltaic/fuel cell/battery energy system in Malaysia hospital. *Energy* **2016**, *112*, 75–90. [CrossRef]
39. Akikur, R.K.; Saidur, R.; Ullah, K.R.; Hajimolana, S.A.; Ping, H.W.; Hussain, M.A. Economic feasibility analysis of a solar energy and solid oxide fuel cell-based cogeneration system in Malaysia. *Clean Technol. Environ. Policy* **2015**, *18*, 669–687. [CrossRef]
40. Guo, X.; Zhang, H.; Zhao, J.; Wang, F.; Wang, J.; Miao, H.; Yuan, J. Performance evaluation of an integrated high-temperature proton exchange membrane fuel cell and absorption cycle system for power and heating/cooling cogeneration. *Energy Convers. Manag.* **2019**, *181*, 292–301. [CrossRef]
41. Di Marcoberardino, G.; Chiarabaglio, L.; Manzolini, G.; Campanari, S. A Techno-economic comparison of micro-cogeneration systems based on polymer electrolyte membrane fuel cell for residential applications. *Appl. Energy* **2019**, *239*, 692–705. [CrossRef]
42. Khani, L.; Mehr, A.S.; Yari, M.; Mahmoudi, S.M.S. Multi-objective optimization of an indirectly integrated solid oxide fuel cell-gas turbine cogeneration system. *Int. J. Hydrog. Energy* **2016**, *41*, 21470–21488. [CrossRef]
43. Asensio, F.J.; San Martín, J.I.; Zamora, I.; Oñederra, O. Model for optimal management of the cooling system of a fuel cell-based combined heat and power system for developing optimization control strategies. *Appl. Energy* **2018**, *211*, 413–430. [CrossRef]
44. Cinti, G.; Bidini, G.; Hemmes, K. Comparison of the solid oxide fuel cell system for micro CHP using natural gas with a system using a mixture of natural gas and hydrogen. *Appl. Energy* **2019**, *238*, 69–77. [CrossRef]
45. Nalbant, Y.; Colpan, C.O.; Devrim, Y. Energy and exergy performance assessments of a high temperature-proton exchange membrane fuel cell based integrated cogeneration system. *Int. J. Hydrog. Energy* **2019**. [CrossRef]
46. Kwan, T.H.; Shen, Y.; Yao, Q. An energy management strategy for supplying combined heat and power by the fuel cell thermoelectric hybrid system. *Appl. Energy* **2019**, *251*. [CrossRef]
47. Nami, H.; Mahmoudi, S.M.S.; Nemati, A. Exergy, economic and environmental impact assessment and optimization of a novel cogeneration system including a gas turbine, a supercritical $CO_2$ and an organic Rankine cycle (GT-HRSG/$SCO_2$). *Appl. Therm. Eng.* **2017**, *110*, 1315–1330. [CrossRef]
48. Hosseinpour, J.; Sadeghi, M.; Chitsaz, A.; Ranjbar, F.; Rosen, M.A. Exergy assessment and optimization of a cogeneration system based on a solid oxide fuel cell integrated with a Stirling engine. *Energy Convers. Manag.* **2017**, *143*, 448–458. [CrossRef]
49. Aviso, K.B.; Tan, R.R. Fuzzy P-graph for optimal synthesis of cogeneration and trigeneration systems. *Energy* **2018**, *154*, 258–268. [CrossRef]

50. Cavallaro, F.; Zavadskas, E.K.; Raslanas, S. Evaluation of Combined Heat and Power (CHP) Systems Using Fuzzy Shannon Entropy and Fuzzy TOPSIS. *Sustainability* **2016**, *8*, 556. [CrossRef]
51. Frangopoulos, C. Effect of reliability considerations on the optimal synthesis, design and operation of a cogeneration system. *Energy* **2004**, *29*, 309–329. [CrossRef]
52. Luo, X.J.; Fong, K.F. Investigation on part-load performances of combined cooling and power system primed by solid oxide fuel cell with different bottoming cycles. *J. Power Sources* **2019**, *429*, 127–178. [CrossRef]
53. Jung, Y.; Kim, J.; Lee, H. Multi-criteria evaluation of medium-sized residential building with micro-CHP system in South Korea. *Energy Build.* **2019**, *193*, 201–215. [CrossRef]
54. Facci, A.L.; Ubertini, S. Analysis of a fuel cell combined heat and power plant under realistic smart management scenarios. *Appl. Energy* **2018**, *216*, 60–72. [CrossRef]
55. Hassanzadeh, H.; Farzad, M.A.; Safavinejad, A.; Agaebrahimi, M.R. Performance assessment of a SOFC cogeneration system for residential buildings located in eastern Iran. *Iran. J. Hydrog. Fuel Cell* **2016**, *2*, 81–97.
56. Vialetto, G.; Rokni, M. A New Cogeneration Residential System Based on Solid Oxide Fuel Cells for a Northern European Climate. In Proceedings of the Global Conference on Global Warming, Athens, Greece, 24–27 May 2015.
57. Wakui, T.; Yokoyama, R. Optimal structural design of residential cogeneration systems with battery based on improved solution method for mixed-integer linear programming. *Energy* **2015**, *84*, 106–120. [CrossRef]
58. Vialetto, G.; Noro, M.; Rokni, M. Combined micro-cogeneration and electric vehicle system for household application: An energy and economic analysis in a Northern European climate. *Int. J. Hydrog. Energy* **2017**, *42*, 10285–10297. [CrossRef]
59. Baldi, F.; Wang, L.; Pérez-Fortes, M.; Maréchal, F. A Cogeneration System Based on Solid Oxide and Proton Exchange Membrane Fuel Cells With Hybrid Storage for Off-Grid Applications. *Front. Energy Res.* **2019**, *6*. [CrossRef]
60. Haghighat Mamaghani, A.; Najafi, B.; Casalegno, A.; Rinaldi, F. Optimization of an HT-PEM fuel cell based residential micro combined heat and power system: A multi-objective approach. *J. Clean. Prod.* **2018**, *180*, 126–138. [CrossRef]
61. Napoli, R.; Gandiglio, M.; Lanzini, A.; Santarelli, M. Techno-economic analysis of PEMFC and SOFC micro-CHP fuel cell systems for the residential sector. *Energy Build.* **2015**, *103*, 131–146. [CrossRef]
62. Rabbani, A.; Rokni, M. Modeling and Analysis of Transport Processes and Efficiency of Combined SOFC and PEMFC Systems. *Energies* **2014**, *7*, 5502–5522. [CrossRef]
63. Agarwal, S.; Chourasiya, S.; Kumawat, R.K.; Palwalia, D.D.K. Performance Analysis of Standalone Hybrid PV- SOFC- BATTERY Generation System. *Natl. Conf. Renew. Energy Environ. (NCREE-2015)* **2015**, *2*, 49–53. [CrossRef]
64. Sadeghi, S.; Ameri, M. Multi-objective optimization of pv-sofc-gt-electrolyser hybrid system. *J. Renew. Energy Environ.* **2015**, *2*, 47–58.
65. Wu, S.; Zhang, H.; Ni, M. Performance assessment of a hybrid system integrating a molten carbonate fuel cell and a thermoelectric generator. *Energy* **2016**, *112*, 520–527. [CrossRef]
66. Sarabchi, N.; Mahmoudi, S.M.S.; Yari, M.; Farzi, A. Exergoeconomic analysis and optimization of a novel hybrid cogeneration system: High-temperature proton exchange membrane fuel cell/Kalina cycle, driven by solar energy. *Energy Convers. Manag.* **2019**, *190*, 14–33. [CrossRef]
67. Löbberding, L.; Madlener, R. Techno-economic analysis of micro fuel cell cogeneration and storage in Germany. *Appl. Energy* **2019**, *235*, 1603–1613. [CrossRef]
68. Jin, Y.; Sun, L.; Shen, J. Thermal economic analysis of hybrid open-cathode hydrogen fuel cell and heat pump cogeneration. *Int. J. Hydrog. Energy* **2019**, *44*, 29692–29699. [CrossRef]
69. Huang, Y.; Wang, W.; Hou, B. A hybrid algorithm for mixed integer nonlinear programming in residential energy management. *J. Clean. Prod.* **2019**, *226*, 940–948. [CrossRef]
70. Bachmann, T.M.; Carnicelli, F.; Preiss, P. Life cycle assessment of domestic fuel cell micro combined heat and power generation: Exploring influential factors. *Int. J. Hydrog. Energy* **2019**, *44*, 3891–3905. [CrossRef]
71. Roshandel, R.; Golzar, F.; Astaneh, M. Technical, economic and environmental optimization of combined heat and power systems based on solid oxide fuel cell for a greenhouse case study. *Energy Convers. Manag.* **2018**, *164*, 144–156. [CrossRef]
72. Romdhane, J.; Louahlia, H.; Marion, M. Dynamic modeling of an eco-neighborhood integrated micro-CHP based on PEMFC: Performance and economic analyses. *Energy Build.* **2018**, *166*, 93–108. [CrossRef]

73. Yoda, M.; Inoue, S.; Takuwa, T.; Yasuhara, K.; Suzuki, M. Development and Commercialization of New Residential SOFC CHP System. *Ecs Trans.* **2017**, *78*, 125–132. [CrossRef]
74. Spazzafumo, G. Cogeneration of power and substitute of natural gas using electrolytic hydrogen, biomass and high temperature fuel cells. *Int. J. Hydrog. Energy* **2018**, *43*, 11811–11819. [CrossRef]
75. Perna, A.; Minutillo, M.; Jannelli, E.; Cigolotti, V.; Nam, S.W.; Yoon, K.J. Performance assessment of a hybrid SOFC/MGT cogeneration power plant fed by syngas from a biomass down-draft gasifier. *Appl. Energy* **2018**, *227*, 80–91. [CrossRef]
76. Giarola, S.; Forte, O.; Lanzini, A.; Gandiglio, M.; Santarelli, M.; Hawkes, A. Techno-economic assessment of biogas-fed solid oxide fuel cell combined heat and power system at industrial scale. *Appl. Energy* **2018**, *211*, 689–704. [CrossRef]
77. Di Marcoberardino, G.; Manzolini, G.; Guignard, C.; Magaud, V. Optimization of a micro-CHP system based on polymer electrolyte membrane fuel cell and membrane reactor from economic and life cycle assessment point of view. *Chem. Eng. Process. Process Intensif.* **2018**, *131*, 70–83. [CrossRef]
78. Chitsaz, A.; Sadeghi, M.; Sadeghi, M.; Ghanbarloo, E. Exergoenvironmental comparison of internal reforming against external reforming in a cogeneration system based on solid oxide fuel cell using an evolutionary algorithm. *Energy* **2018**, *144*, 420–431. [CrossRef]
79. Budak, Y.; Devrim, Y. Investigation of micro-combined heat and power application of PEM fuel cell systems. *Energy Convers. Manag.* **2018**, *160*, 486–494. [CrossRef]
80. Tanaka, T.; Kamiko, H.; Akiba, K.; Ito, S.; Osaki, H.; Yashiro, M.; Inui, Y. Energetic analyses of installing SOFC co-generation systems with EV charging equipment in Japanese cafeteria. *Energy Convers. Manag.* **2017**, *153*, 435–445. [CrossRef]
81. Mehr, A.S.; Gandiglio, M.; MosayebNezhad, M.; Lanzini, A.; Mahmoudi, S.M.S.; Yari, M.; Santarelli, M. Solar-assisted integrated biogas solid oxide fuel cell (SOFC) installation in wastewater treatment plant: Energy and economic analysis. *Appl. Energy* **2017**, *191*, 620–638. [CrossRef]
82. Kupecki, J.; Skrzypkiewicz, M.; Wierzbicki, M.; Stepien, M. Experimental and numerical analysis of a serial connection of two SOFC stacks in a micro-CHP system fed by biogas. *Int. J. Hydrog. Energy* **2017**, *42*, 3487–3497. [CrossRef]
83. Karami, H.; Sanjari, M.J.; Gooi, H.B.; Gharehpetian, G.B.; Guerrero, J.M. Stochastic analysis of residential micro combined heat and power system. *Energy Convers. Manag.* **2017**, *138*, 190–198. [CrossRef]
84. Hormaza-Mejia, A.; Zhao, L.; Brouwer, J. SOFC Micro-CHP system with thermal energy storage in residential applications. In Proceedings of the ASME 2017 15th International Conference on Fuel Cell Science, Engineering and Technology FUELCELL2017, Charlotte, NC, USA, 26–30 June 2017; pp. 1–6.
85. Hajabdollahi, Z.; Fu, P.-F. Multi-objective based configuration optimization of SOFC-GT cogeneration plant. *Appl. Therm. Eng.* **2017**, *112*, 549–559. [CrossRef]
86. Haghighat Mamaghani, A.; Najafi, B.; Casalegno, A.; Rinaldi, F. Predictive modelling and adaptive long-term performance optimization of an HT-PEM fuel cell based micro combined heat and power (CHP) plant. *Appl. Energy* **2017**, *192*, 519–529. [CrossRef]
87. Eveloy, V.; Rodgers, P.; Al Alili, A. Multi-objective optimization of a pressurized solid oxide fuel cell–gas turbine hybrid system integrated with seawater reverse osmosis. *Energy* **2017**, *123*, 594–614. [CrossRef]
88. Anyenya, G.A.; Sullivan, N.P.; Braun, R.J. Modeling and simulation of a novel 4.5 kW e multi-stack solid-oxide fuel cell prototype assembly for combined heat and power. *Energy Convers. Manag.* **2017**, *140*, 247–259. [CrossRef]
89. Zhang, X.; Ni, M.; Dong, F.; He, W.; Chen, B.; Xu, H. Thermodynamic analysis and performance optimization of solid oxide fuel cell and refrigerator hybrid system based on $H_2$ and CO. *Appl. Therm. Eng.* **2016**, *108*, 347–352. [CrossRef]
90. Reyhani, H.A.; Meratizaman, M.; Ebrahimi, A.; Pourali, O.; Amidpour, M. Thermodynamic and economic optimization of SOFC-GT and its cogeneration opportunities using generated syngas from heavy fuel oil gasification. *Energy* **2016**, *107*, 141–164. [CrossRef]
91. Pohl, E.; Meier, P.; Maximini, M.; Schloß, J.v. Primary energy savings of a modular combined heat and power plant based on high temperature proton exchange membrane fuel cells. *Appl. Therm. Eng.* **2016**, *104*, 54–63. [CrossRef]

92. Misra, S.; Satyaprasad, G.; Mahapatra, S.S.; Mohanty, A.; Biswal, S.R.; Dora, A. Design of Fuel cell based Co-generation systems: An approach for battery less Solar PV system. *Int. Res. J. Eng. Technol.* **2016**, *3*, 47–50.
93. Khani, L.; Mahmoudi, S.M.S.; Chitsaz, A.; Rosen, M.A. Energy and exergoeconomic evaluation of a new power/cooling cogeneration system based on a solid oxide fuel cell. *Energy* **2016**, *94*, 64–77. [CrossRef]
94. Fong, K.F.; Lee, C.K. System analysis and appraisal of SOFC-primed micro cogeneration for residential application in subtropical region. *Energy Build.* **2016**, *128*, 819–826. [CrossRef]
95. Assaf, J.; Shabani, B. Transient simulation modelling and energy performance of a standalone solar-hydrogen combined heat and power system integrated with solar-thermal collectors. *Appl. Energy* **2016**, *178*, 66–77. [CrossRef]
96. Windeknecht, M.; Tzscheutschler, P. Optimization of the Heat Output of High Temperature Fuel Cell Micro-CHP in Single Family Homes. *Energy Procedia* **2015**, *78*, 2160–2165. [CrossRef]
97. Vialetto, G.; Rokni, M. Innovative household systems based on solid oxide fuel cells for a northern European climate. *Renew. Energy* **2015**, *78*, 146–156. [CrossRef]
98. Ullah, K.R.; Akikur, R.K.; Ping, H.W.; Saidur, R.; Hajimolana, S.A.; Hussain, M.A. An experimental investigation on a single tubular SOFC for renewable energy based cogeneration system. *Energy Convers. Manag.* **2015**, *94*, 139–149. [CrossRef]
99. Shariatzadeh, J.O.; Refahi, A.H.; Abolhassani, S.S.; Rahmani, M. Modeling and optimization of a novel solar chimney cogeneration power plant combined with solid oxide electrolysis/fuel cell. *Energy Convers. Manag.* **2015**, *105*, 423–432. [CrossRef]
100. Liso, V.; Zhao, Y.; Yang, W.; Nielsen, M. Modelling of a Solid Oxide Fuel Cell CHP System Coupled with a Hot Water Storage Tank for a Single Household. *Energies* **2015**, *8*, 2211–2229. [CrossRef]
101. Kupecki, J.; Jewulski, J.; Motylinski, K. Parametric evaluation of a micro-CHP unit with solid oxide fuel cells integrated with oxygen transport membranes. *Int. J. Hydrog. Energy* **2015**, *40*, 11633–11640. [CrossRef]
102. Ham, S.-W.; Jo, S.-Y.; Dong, H.-W.; Jeong, J.-W. A simplified PEM fuel cell model for building cogeneration applications. *Energy Build.* **2015**, *107*, 213–225. [CrossRef]
103. Elmer, T.; Worall, M.; Wu, S.; Riffat, S.B. Emission and economic performance assessment of a solid oxide fuel cell micro-combined heat and power system in a domestic building. *Appl. Therm. Eng.* **2015**, *90*, 1082–1089. [CrossRef]
104. Cappa, F.; Facci, A.L.; Ubertini, S. Proton exchange membrane fuel cell for cooperating households: A convenient combined heat and power solution for residential applications. *Energy* **2015**, *90*, 1229–1238. [CrossRef]
105. Canelli, M.; Entchev, E.; Sasso, M.; Yang, L.; Ghorab, M. Dynamic simulations of hybrid energy systems in load sharing application. *Appl. Therm. Eng.* **2015**, *78*, 315–325. [CrossRef]
106. Borji, M.; Atashkari, K.; Ghorbani, S.; Nariman-Zadeh, N. Parametric analysis and Pareto optimization of an integrated autothermal biomass gasification, solid oxide fuel cell and micro gas turbine CHP system. *Int. J. Hydrog. Energy* **2015**, *40*, 14202–14223. [CrossRef]
107. Arsalis, A.; Kær, S.K.; Nielsen, M.P. Modeling and optimization of a heat-pump-assisted high temperature proton exchange membrane fuel cell micro-combined-heat-and-power system for residential applications. *Appl. Energy* **2015**, *147*, 569–581. [CrossRef]
108. Antonucci, V.; Brunaccini, G.; De Pascale, A.; Ferraro, M.; Melino, F.; Orlandini, V.; Sergi, F. Integration of µ-SOFC Generator and ZEBRA Batteries for Domestic Application and Comparison with other µ-CHP Technologies. *Energy Procedia* **2015**, *75*, 999–1004. [CrossRef]
109. Worall, M.; Elmer, T.; Riffat, S.; Wu, S.; Du, S. An experimental investigation of a micro-tubular SOFC membrane-separated liquid desiccant dehumidification and cooling tri-generation system. *Appl. Therm. Eng.* **2017**, *120*, 64–73. [CrossRef]
110. Elmer, T.; Worall, M.; Wu, S.; Riffat, S. Assessment of a novel solid oxide fuel cell tri-generation system for building applications. *Energy Convers. Manag.* **2016**, *124*, 29–41. [CrossRef]
111. Zhang, S.-L.; Wang, H.; Lu, M.Y.; Zhang, A.-P.; Mogni, L.V.; Liu, Q.; Li, C.-X.; Li, C.-J.; Barnett, S.A. Cobalt-substituted SrTi 0.3 Fe 0.7 O 3−δ: A stable high-performance oxygen electrode material for intermediate-temperature solid oxide electrochemical cells. *J. Energy Environ. Sci. Technol.* **2018**, *11*, 1870–1879. [CrossRef]

112. Chen, T.; Liu, S.; Zhang, J.; Tang, M. Study on the characteristics of GDL with different PTFE content and its effect on the performance of PEMFC. *J Int. J. Heat Mass Transf.* **2019**, *128*, 1168–1174. [CrossRef]
113. Soloveichik, G.L. Flow batteries: Current status and trends. *Chem. Rev.* **2015**, *115*, 11533–11558. [CrossRef]

*Article*

# Modeling, Management, and Control of an Autonomous Wind/Fuel Cell Micro-Grid System

**Ibrahem E. Atawi [1], Ahmed M. Kassem [2] and Sherif A. Zaid [3,*]**

1 Department of Electrical Engineering, Faculty of Engineering, University of Tabuk, Tabuk 71491, Saudi Arabia; ieatawi@ut.edu.sa
2 Department of Electrical Engineering, Faculty of Engineering, Sohag University, Sohag 82524, Egypt; kassem_ahmed53@hotmail.com
3 Department of Electrical Power, Faculty of Engineering, Cairo University, Cairo 12613, Egypt
* Correspondence: shfaraj@ut.edu.sa or sherifzaid3@yahoo.com; Tel.: +2-023-567-8278

Received: 3 January 2019; Accepted: 4 February 2019; Published: 8 February 2019

**Abstract:** This paper proposes a microelectric power grid that includes wind and fuel cell power generation units, as well as a water electrolyzer for producing hydrogen gas. The grid is loaded by an induction motor (IM) as a dynamic load and constant impedance load. An optimal control algorithm using the Mine Blast Algorithm (MBA) is designed to improve the performance of the proposed renewable energy system. Normally, wind power is adapted to feed the loads at normal circumstances. Nevertheless, the fuel cell compensates extra load power demand. An optimal controller is applied to regulate the load voltage and frequency of the main power inverter. Also, optimal vector control is applied to the IM speed control. The response of the microgrid with the proposed optimal control is obtained under step variation in wind speed, load impedance, IM rotor speed, and motor mechanical load torque. The simulation results indicate that the proposed renewable generation system supplies the system loads perfectly and keeps up the desired load demand. Furthermore, the IM speed performance is acceptable under turbulent wind speed.

**Keywords:** wind energy; fuel cell; IM; induction generator; hybrid system; mine blast optimizer

## 1. Introduction

Modern industries, transportation means, and nearly all mankind's requirements mainly depend on electrical power. Traditionally, electrical power generation is essentially based on fossil fuel resources. Nevertheless, fossil fuels suffer from several drawbacks, such as depletion by 2050. However, the rapid growth of the world's population increases the world electrical power demand. The global energy demand estimated 2.1% in 2017 (more than twice the average increase over the previous five years) [1]. Also, it generates harmful emissions that form the essential cause of the phenomena of global warming and many environmental problems. Energy-related carbon dioxide ($CO_2$) emissions rose—by an estimated 1.4% in 2017—for the first time in four years, at a time when climate scientists said that emissions needed to be in steep decline [1]. Several decades ago, renewable energy resources have gained more attention as a sustainable replacement for fossil fuels. Renewable energy resources have great advantages as they are clean, do not deplete, and are available everywhere. Many renewable energy resources [2–5] have been introduced recently, such as photovoltaic (PV), wind, ocean wave, ocean tides, and micro hydro. Thanks to technology advances and rapid growth, dramatic reductions in the costs of solar PV and wind energy systems have occurred [6]. In the same context, new energy alternative technologies like biomass, geothermal, microturbines, and fuel cells (FCs) have been investigated [7]. Now, electricity generation by renewable systems is less expensive than the newly installed fossil and nuclear power plants in many parts of the world. An assessment of different renewable energy resources for electrical power production showed that wind energy is the first choice [8].

Renewable energy systems may be classified into grid-connected systems and standalone systems. The present capacities of the grid-connected renewable energy systems vary from several kilowatts of residential PV systems to large-scale wind farms. The grid-connected systems do not need any storage as the generated energy is injected directly to the grid. These systems are suitable for urban regions where the grid is available. However, standalone systems are suitable for rural areas, where grid extension is not feasible. In standalone renewable energy systems, the load is an individual house and not connected to a grid. The capacities of these systems are usually small. In some applications, several houses are connected to form a small power grid called microgrids (MGs) [9,10]. Microgrid technology has become popular in islands as it provides a cost-effective alternative where power grid extension is expensive and fuel transportation is difficult and costly [11,12].

The major obstacle for utilizing one technology of renewable energy sources is the intermittent nature of that source. That intermittent behavior of the renewable energy sources comes from the strong dependency on the environmental conditions, which are changing continuously. A suggestion to solve the intermittency problem of the renewable energy systems is the use of energy storage element. Energy storage units are classified as capacity-oriented storage systems and access-oriented storage systems. The capacity-oriented storage systems include pumped hydroelectric storage, compressed air energy storage, and hydrogen storage systems. It has a slow response and is considered long-term energy storage. Batteries, superconducting magnetic energy storage, supercapacitors, and flywheels are considered access-oriented storage systems. It has a fast time response that is useful for short duration disturbance applications [13–16]. In fact, the integration of energy storage systems with one technology of renewable energy sources has many disadvantages. One of them is the load power variations, which may harm the storage system and degrade its lifetime. On the other hand, the size and cost of the system increase [17]. Hybrid power generation systems are introduced essentially to alleviate these disadvantages. These systems contain two or more energy sources with a storage system. Hybrid renewable energy systems have benefits of high reliability, high efficiency, better power quality, and low energy storage requirements [18,19].

Usually, microgrids can operate in two modes: grid-connected mode and autonomous mode. So, the main benefit of a microgrid is that it is able to operate in both the above two modes [20]. The microgrid can then function autonomously. Both loads and generation in microgrids are usually interconnected at low levels of voltages. However, one issue regarding the microgrid is that the operator needs to be very vigilant because numbers of power system areas are connected to microgrid. Also, in microgrid generation resources can include wind, photovoltaic, fuel cells, energy storage, or other power generation sources [21].

In the literature, there are different types of standalone hybrid power sources have been reported [22–27]. They usually combine solar energy and/or wind energy with another green power source such as FC, biomass, etc. As the work in this paper is directed to hybrid wind/FC generation systems, we focus on the literature review of that subject. Khan et al. [28] investigates a hybrid wind/FC generation system and presented a detailed life cycle analysis of the system for application in Newfoundland and Labrador. It concludes that the system is highly nonlinear and difficult to model. Battista et al. [29] presents a wind/hydrogen power system with a novel power conditioning algorithm. It introduces a Maximum Power Point Tracking (MPPT) control strategy that was developed using concepts of the reference conditioning technique and of the sliding mode control theory. Gorgun et al. [30] presents a wind/hydrogen power system and developed the electrolyzer and hydrogen storage dynamic model. Bizon et al. [31] introduces a hybrid wind/FC generation system with a global extremum seeking a control algorithm for optimal operation of the wind turbine under the turbulent wind.

This paper presents a novel hybrid wind/FC energy system. In addition, it presents the system detailed dynamic model, the application of mine blast optimization, the controller design, the performance analyses, and the simulations of the developed system under the turbulent wind speed.

In this study, the microelectrical power grid is managed and controlled based on an optimal controller using mine blast algorithm. The proposed microgrid system mainly consists of R-L static load, IM as a dynamic load, a wind generation system, FC generation system, water electrolyzer, uncontrolled rectifier, and controlled DC/AC converter.

The power system management is adapted to make the wind power generation is the master source for the loads. Also, hydrogen gas is produced using a water electrolyzer during wind power generation peaks. Hydrogen is fed back to the FC system adding to its energy storage.

The main contribution of this study can be seen from the obtained results that the proposed renewable generation system can supply the system loads perfectly in addition to a better prediction of the electrical parameter waveforms. Also, the controller response follows up the desired load demand with a small maximum overshoot and little settling time. In addition to, the generated power is managed such that the load required power is supplied by the wind power and the more needed power is covered by the fuel cell generation unit.

This paper is arranged as follows. Section 2 includes the discussion of the system description. Section 3 presents the proposed system dynamic model. Details of the system controllers are found in Section 4. The simulation results and their discussion are presented in Section 5. Finally, Section 6 shows the conclusions.

## 2. System Description

The proposed system is a hybrid wind/FC microgrid supplies two loads, as shown in Figure 1. The wind turbine drives a 3-φ Induction Generator (IG) The output voltage of the IG is rectified via a diode rectifier producing the Direct Current (DC) bus voltage. That bus supplies a water electrolyzer that produces hydrogen gas to be saved in the FC generator. Then, the output voltage of the FC is connected to the system DC bus. In addition, the DC bus supplies the 3-φ inverter that converts DC power into Alternating Current (AC) power to feed the system loads. General R-L impedance represents the static load. In addition, the dynamic load is a speed controlled induction motor. That inverter is controlled in such a way to supply loads with a regulated AC voltage and frequency.

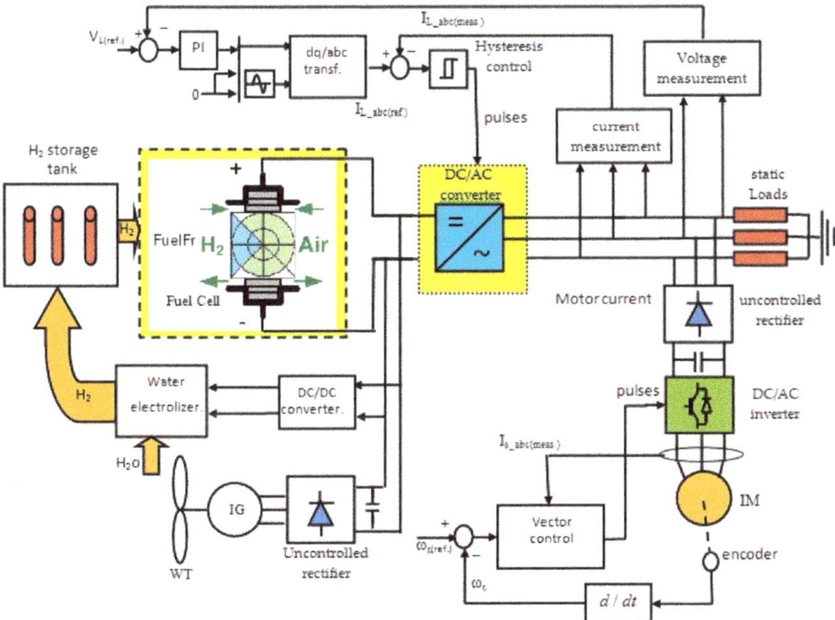


**Figure 1.** The proposed microgrid energy system and its controllers.

Usually, wind speed variations cause the IG output power to vary as well. Therefore, an FC unit supplies power to loads when the power of the wind generation unit drops. Hence, it acts as a slave to compensate for any decrease in the generated wind energy. In addition, it can supply additional power demanded by the loads.

Frequently, the load voltage of the system is not regulated due to the speed variations of the wind and load changes. Therefore, the voltage controller is used for the main inverter to regulate the load voltage and frequency. Also, the optimization algorithm is applied to control the speed of a vector controlled induction motor.

## 3. The Dynamic Model of the System

The detailed dynamic model of all proposed system components will be discussed in the following paragraphs.

### 3.1. Wind Turbine Model

The power output of the wind turbine can be represented as [32]

$$P_m = 0.5C_p(\lambda, \beta)A\rho v_w^3 \tag{1}$$

where, $A = \pi R^2$ is the swept area ($m^2$) by the blades, $\beta$ is the blade pitch angle (in degrees), $R$ is the radius of the turbine blade, $v_w$ is the wind speed, $\rho$ is the air density ($kg/m^3$), $\lambda$ is the tip–speed ratio as defined by Equation (2), $C_p$ is the performance coefficient of the turbine that is given by Equation (3), and $\omega_m$ is the turbine mechanical angular rotor speed.

$$\lambda = R\omega_m / v_w \tag{2}$$

$$C_p(\lambda, \beta) = (0.44 - 0.0167\beta)\sin\pi\left(\frac{\lambda - 3}{15 - 0.3\beta}\right) - 0.00184(\lambda - 3)\beta \tag{3}$$

The wind turbine torque ($T_m$) is related to its power by the classical relation:

$$T_m = P_m / \omega_m \tag{4}$$

The dynamic equation of the wind turbine and the electrical generator mechanical system can be written as

$$J\frac{d\omega_m}{dt} = T_m - T_e - B\omega_m \tag{5}$$

where, $T_e$ is the electrical generator electromagnetic torque (N·m), $J$ is the combined inertia of the generator rotor and the wind turbine (kg·m), and $B$ is the mechanical viscous friction (N·m·s/rad).

### 3.2. Dynamic Model of the Induction Machine

In a synchronous reference frame, the induction motor dynamic model may be represented by [33]

$$\frac{di_{ds}}{dt} = \frac{1}{\sigma L_s}\left[-\left(r_s + \frac{r_r L_m^2}{L_r^2}\right)i_{ds} + \omega_s \sigma L_s i_{qs} + \frac{r_r L_m}{L_r}\lambda_{dr} + \frac{L_m}{L_r}\omega_r \lambda_{qr} + v_{ds}\right] \tag{6}$$

$$\frac{di_{qs}}{dt} = \frac{1}{\sigma L_s}\left[-\left(r_s + \frac{r_r L_m^2}{L_r^2}\right)i_{qs} - \omega_s \sigma L_s i_{ds} + \frac{r_r L_m}{L_r}\lambda_{qr} - \frac{L_m}{L_r}\omega_r \lambda_{dr} + v_{qs}\right] \tag{7}$$

$$\frac{d\lambda_{dr}}{dt} = (\omega_s - \omega_r)\lambda_{qr} + \frac{r_r L_m}{L_r}i_{ds} - \frac{r_r}{L_r}\lambda_{dr} \tag{8}$$

$$\frac{d\lambda_{qr}}{dt} = -(\omega_s - \omega_r)\lambda_{dr} + \frac{r_r L_m}{L_r}i_{qs} - \frac{r_r}{L_r}\lambda_{qr} \tag{9}$$

$$J_m \frac{d\omega_r}{dt} = T_l - T_{em} - B_m \omega_r \tag{10}$$

where, $r_r$ is the rotor resistance; ($v_{ds}$ and $v_{qs}$) are the stator d- and q-axis voltage components, respectively; ($i_{ds}$ and $i_{qs}$) are the stator d- and q-axis current components, respectively; ($\lambda_{dr}$ and $\lambda_{qr}$) are the rotor d- and q-axis flux linkage components, respectively; ($L_s$, $L_r$, and $L_m$) are the stator inductance, rotor, and mutual inductances, respectively; $\omega_r$ is the motor speed; $\omega_s$ is the motor synchronous speed; $T_{em}$ is the motor electromagnetic torque (N·m); $J_m$ is the inertia of the IM rotor (kg·m); and $B_m$ is the viscous friction of the coupling (N·m·s/rad).

### 3.3. Fuel Cell Model

Typically, the Proton-Exchange Membrane Fuel Cell (PEM FC) has an electrical characteristic at normal environmental conditions as shown in Figure 2. Normally, FCs are subjected to internal voltage losses that cause a voltage drop beyond the nominal voltage values. Three kinds of voltage losses are presented: the ohmic polarization, the concentration polarization, and the activation polarization. The slowness of the chemical reactions is the cause behind the cell activation losses; it can be reduced by maximizing the catalyst contact area. However, the cause of the resistive losses is the resistance of all the FC electrical circuit and their connections. This part of losses can be alleviated by well hydrating the membrane. Finally, the concentration losses come from the changes in gas concentration at the electrodes surface.

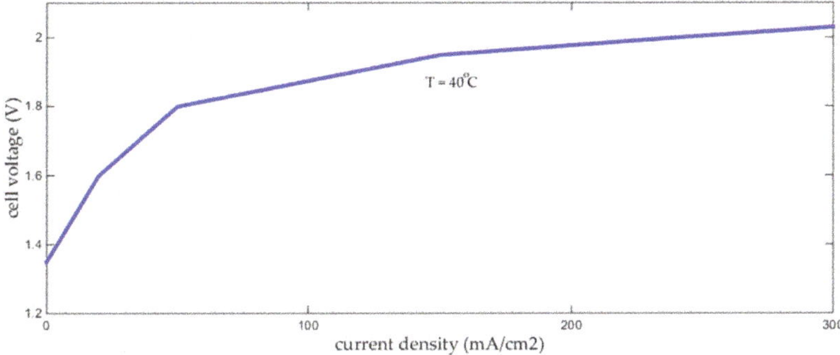


**Figure 2.** The V–I characteristic of the Proton-Exchange Membrane (PEM) fuel cell at normal environmental conditions.

The model of the PEM FC is given by [34–36]

$$E = N\left[E_o + \frac{R'T}{nF} Ln\left(\frac{P_{H2}\left(\frac{P_{O2}}{P_{std}}\right)}{P_{H2Oc}}\right) - V_{drop}\right] \tag{11}$$

$$V_{drop} = \frac{R'T}{nF}\left[Ln\left(\frac{i_n + i}{i_o}\right) + \frac{nF\alpha}{R'T}(i_n + i) - Ln\left(1 - \frac{i_n + i}{i_L}\right)\right] \tag{12}$$

where, $E$ is the stack output voltage, $E_o$ is the cell open circuit voltage at standard pressure, $N$ is the number of cells in stack, $F$ is Faraday's constant, $n$ is the number of transferred electrons in the electrochemical reaction, $R'$ is the universal gas constant, $T$ is the operating temperature, $P_{H2}$ is the partial pressure of hydrogen, $P_{O2}$ is the partial pressure of oxygen, $P_{H2Oc}$ is the partial pressure of gas water, $P_{std}$ is the standard pressure, $V_{drop}$ is the voltage losses, $i$ is the output current density, $i_n$ is the internal current density related to internal current losses, $i_o$ is the exchange current density related to activation losses, $i_L$ is the limiting current density related to concentration losses, and $\alpha$ is the area specific resistance related to resistive losses.

### 3.4. Uncontrolled Rectifier Model

The IG speed is directly related to the wind speed that changes usually with time. Hence, the IG output voltage is not regulated in terms of its magnitude or frequency. This issue is not suitable for many applications that require regulated sources. That problem can be alleviated by rectifying the IG output voltage to form the DC bus then converting it to an AC voltage via a power inverter. The rectifier is simply a diode bridge rectifier. Neglecting the source inductance, the average model of the rectifier is given by [37]

$$V_d = 3\sqrt{3}/\pi V_g \quad , \quad I_d = \pi/2\sqrt{3}I_g \tag{13}$$

where, ($I_g$, $V_g$) are the phase RMS current and voltage of the IG, respectively, and ($I_d$, $V_d$) are the average rectifier output current and voltage, respectively.

### 3.5. Boost Converter Model

The classical circuit diagram of the boost converter is shown in Figure 3. The input of the boost converter is the DC bus voltage. However, its output feeds the water electrolyzer. Its function is to regulate the power transfer to the water electrolyzer in turn to the fuel cell. The average model of the boost converter is given by [38]

$$V_d = V_{fc}/(1-d) \quad , \quad I_d = (1-d)I_{fc} \tag{14}$$

where, $d$ is the duty ratio of the switch and ($V_{fc}$, $I_{fc}$) are the fuel cell output voltage and current, respectively.

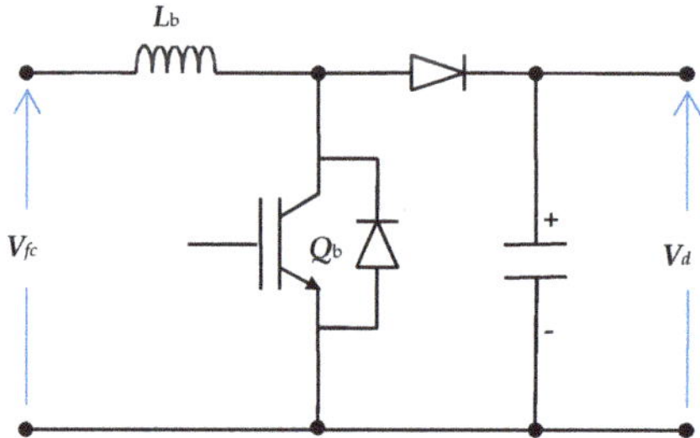


**Figure 3.** The boost circuit diagram.

### 3.6. Main Power Inverter Model

The power circuit diagram of a 3-φ inverter connected to L-C filter is shown in Figure 4a. The output voltage $\underline{V}_c$, the inverter voltage $\underline{V}_i$, the output current $\underline{I}_o$, and the filter current $\underline{I}_f$ are expressed as space vectors by

$$\underline{F} = 2/3\left(f_a + af_b + a^2 f_c\right) \tag{15}$$

where, ($f_a$, $f_b$, and $f_c$) are the phase values, $\underline{F}$ is the space vector of the quantity, and $a = e^{j(2\pi/3)}$.

The switching states of the inverter are determined by its gate signals ($S_a$, $S_b$, and $S_c$). These states can also be expressed as space vector $\underline{S}$ using Equation (13). Considering the possible combinations of the gate signals, there are eight switching states. These states generate eight voltage vectors as shown in Figure 4b. There are two zero voltage vectors ($\underline{V}_0 = \underline{V}_7$) and six active voltage vectors. The system dynamic behavior can be expressed by

$$L\frac{d\underline{I}_f}{dt} = \underline{V}_i - \underline{V}_c \tag{16}$$

$$C\frac{d\underline{V}_c}{dt} = \underline{I}_f - \underline{I}_o \tag{17}$$

$$\underline{V}_i = V_d\underline{S} \tag{18}$$

where, ($L$, $C$) are the filter inductance and capacitance, respectively.

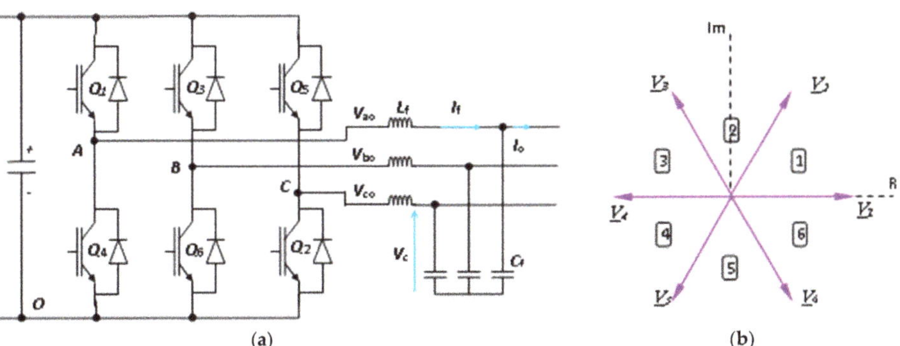

**Figure 4.** (**a**) The power circuit diagram of a 3-φ inverter. (**b**) The 3-φ inverter space vectors.

## 4. System Controllers

The control system of the proposed wind/FC system consists of three controllers. The first controller is the main inverter controller that regulates the load voltage and frequency. The second controller is the boost converter controller. However, the third controller is the induction motor controller that controls the speed of the induction motor. The three controllers will be discussed in the following paragraphs.

### 4.1. Main Inverter Controller

The proposed controller is shown in Figure 5, a current controlled voltage source inverter (VSI) is employed. The inverter output voltage is compared to the reference voltage generating an error signal. The error is sent to an optimized Proportional-Integral (PI) controller that generates the reference three-phase currents. These reference currents are compared to the actual three-phase currents producing error signals that are fed to hysteresis controllers to produce the inverter switches driving pulses.

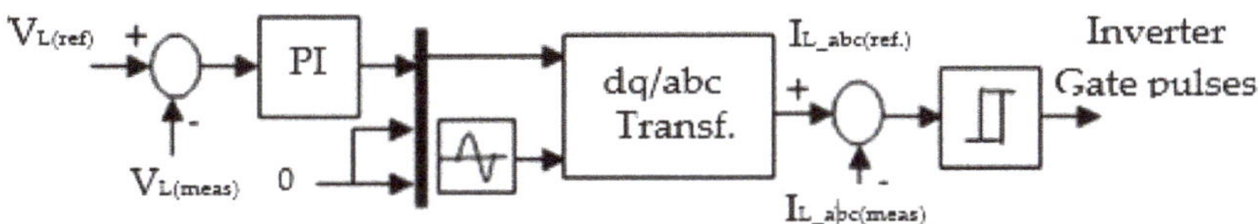


**Figure 5.** Block diagram of the inverter controller.

### 4.2. Mine Blast Optimization Algorithm

The idea behind this algorithm is the exploration technique of landmines. An initial shot point ($z_o$) is adapted using [39,40]

$$\vec{z}_o = SB + \{rand\} \times \left\{ \vec{LB} - \vec{SB} \right\} \;, 0 < rand < 1 \tag{19}$$

where, $z_o$ is the first shot point and ($SB$ and $LB$) are the problem upper and lower limits, respectively.

Assume that the population has $N_s$ Shrapnel pieces. Mine blast algorithm has two phases named exploitation and exploration. The function of the exploitation phase is to encourage and to converge the solution. On the other hand, the exploration has the responsibility of exploring the search space.

During the starting iterations of MBA, the exploration factor ($\gamma$) explores the search spaces then check the number of iterations ($i$). The exploration phase ends when ($\gamma$ ) is greater than ($i$) which is given by [40]

$$\vec{z}_{e(i)} = \left\{ \vec{d}_{i-1} \right\} \times (|randn|)^2 \times \cos\left(\frac{360}{N_s}\right) \quad i = 1,2,\ldots,\gamma \tag{20}$$

Hence, the directions of the shrapnel pieces are given by

$$m_{(i)} = \frac{F_{(i)} - F_{(i-1)}}{\vec{z}_{e(i)} - \vec{z}_{e(i-1)}} \quad i = 1,\ 2,\ldots,\gamma \tag{21}$$

The best locations of the shrapnel pieces are calculated by

$$\vec{z}_{(i)} = z_{e(i)} + exp\left(-\sqrt{\frac{\vec{m}_i}{\vec{d}_i}}\right) \times \vec{z}_{e(i)} \quad i = 1,\ 2,\ldots,\gamma \tag{22}$$

where $\vec{d}_{i-1}$ is the shrapnel distance of the exploded mines, $F$ is the fitness function, and $\vec{z}_e$ is the best location.

Exploitation phase can be defined as

$$d_i = \sqrt{\left(\vec{z}_{e(i)} - \vec{z}_{e(i-1)}\right)^2 + \left(F_{(i)} - F_{(i-1)}\right)^2}, \quad i = \gamma + 1,\ldots, Max\_iteration \tag{23}$$

$$\vec{z}_{e(i)} = \left\{ \vec{d}_{i-1} \right\} \times \{rand\} \times cos\alpha \quad i = \gamma + 1,\ldots, Max\_iteration \tag{24}$$

The initial distances of the shrapnel pieces are gradually reduced in the exploitation phase. This can be achieved by reducing the user to converge constant ($\sigma$). The reduction in the initial distance is determined using

$$\vec{d}_i = \frac{\vec{d}_{i-1}}{e^{\left(\frac{i}{\sigma}\right)}} \quad i = 1,2,\ldots, Max\_iteration \tag{25}$$

The mine blast algorithm may be summarized in the flowchart of Figure 6.

Usually, MBA uses an objective function check the optimality of the resulted parameters. There are several forms of the objective function [41], such as the Integral Time Absolute Error (ITAE), Integral Square Error (ISE), and Integral Time Square Error (ITSE). Nevertheless, ITAE is selected due to its better performance. Therefore, the suggested objective function in this paper is ITAE that is given by

$$ITAE = \int_0^{t_s} ((|\Delta e_1| + |\Delta e_2|) \times t)\ dt \tag{26}$$

$$\Delta e_1 = \Delta\omega_{r(ref)} - \Delta\omega_r \tag{27}$$

$$\Delta e_2 = \Delta V_{Load(ref)} - \Delta V_{Load} \tag{28}$$

where, $t_s$ is the simulation time and [$K_{p1}$, $K_{i1}$, $K_{p2}$, and $K_{i2}$] are the parameters to be estimated, $x$ =, and the constraints are assumed to be

$$0.45 \le K_{pz} < 15\ ,\ 0.45 \le K_{iz} < 15 \qquad z = 1,\ 2 \tag{29}$$

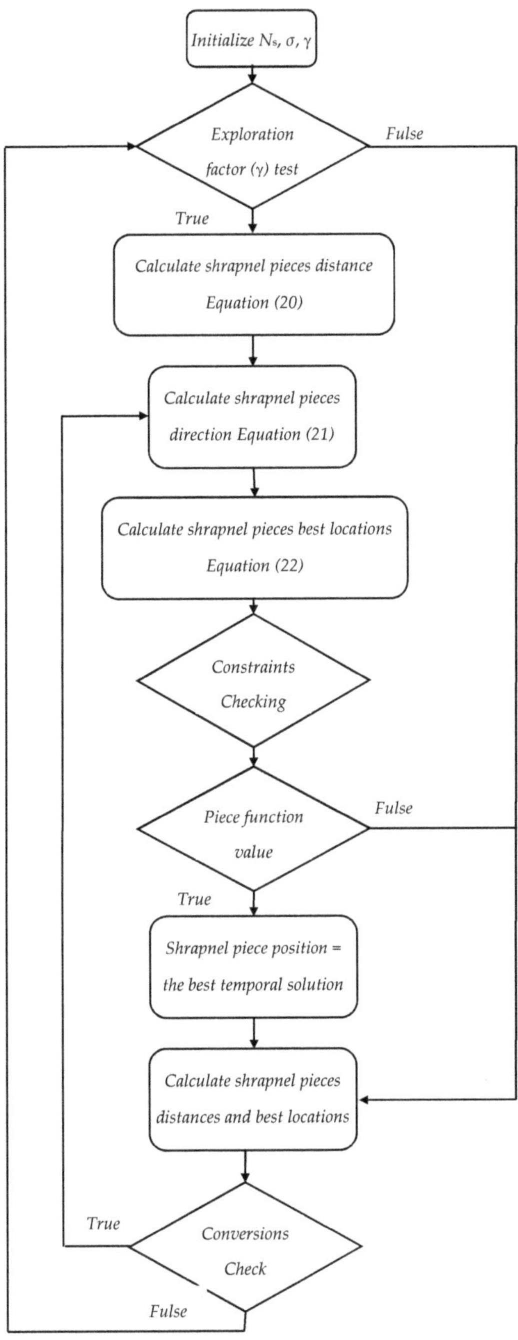


**Figure 6.** Mine blast algorithm flowchart.

The proposed optimal controlling parameters of MBA are given in Table 1.

**Table 1.** Optimal mine blast algorithm controlling parameters.

| Parameter | Value |
|---|---|
| $\alpha$ | 642.3 |
| $N_s$ | 100 |
| No. of variables | 2 |
| Max_iteration | 100 |
| Final distance | 0.0042351 |
| Number of function evaluations | 10,000 |

### 4.3. Boost Converter Controller

The boost converter control circuit is shown in Figure 7. It is a simple voltage regulator. The reference voltage signal is generated based on the voltage error. Then the error is fed to the Proportional-Integral-Differentiator (PID) controller that generates the modulating signal to the Pulse Width Modulator (PWM) unit. In turn, the PWM unit generates the suitable duty cycle pulses for the converter switch.

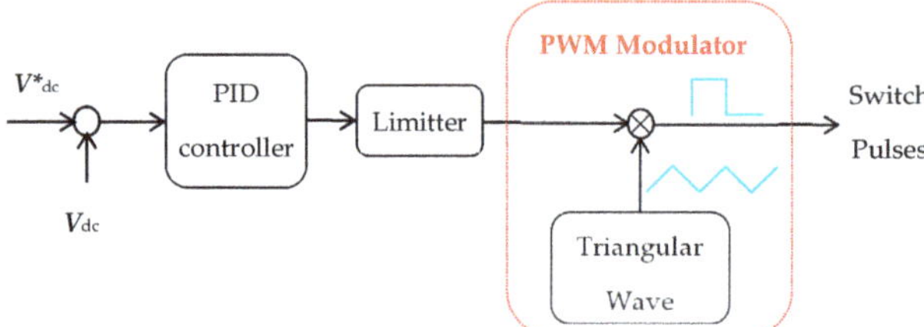


**Figure 7.** Boost converter controller block diagram.

### 4.4. Induction Motor Controller

There are two techniques to control IM: vector and scalar control. The vector control technique is precise and has a high-performance operation. Hence, the speed of the induction motor is controlled using an optimal vector control technique as shown in Figure 8. The actual speed is measured and compared to the reference speed producing the error that is manipulated by an optimal PI controller. The controller generates the reference torque for the vector control. The rotor flux and torque of the IM are estimated using the IM model. The details of vector control and estimators are presented by Trzynadlowski et al. [42].

## 5. Simulation Results

Computer simulations have been carried out to prove the performance of the proposed system under loads and wind speed changes. The proposed system shown in Figure 2 is simulated using MATLAB software package (MATLAB 16, Math Works, Torrance, CA, USA) and tested under various values of wind velocity, IM speed changes, and static load changes. The management of the energy exchange algorithm of the proposed island microgrid is shown by the flowchart of Figure 9.

Figure 10 shows the obtained results for various system parameters like wind speed, FC power, IG (torque-stator-current-speed), the FC pressure of hydrogen and oxygen, the power required by the load, Dc bus voltage, static load (current–voltage), and IM (speed-torque-stator current). The system response is tested at step changes in wind velocity, load impedance, IM speed, and IM load torque. This figure shows that the wind speed varies between 11 and 14 m/s, as shown in Figure 10a. This figure shows also that as the wind speed increases the IG (speed-torque-stator current) increases as well, as indicated in Figure 10b–d, respectively. Enlarging of Figure 10d is shown in Figure 10e. The FC pressure of $H_2$ and $O_2$ are present in Figure 10f. Also, the wind power increases with the wind speed increase as shown in Figure 10g. On the other hand, Figure 10g shows that the fuel cell compensates any reduction in wind power and the load power is the sum of the wind power plus FC power.

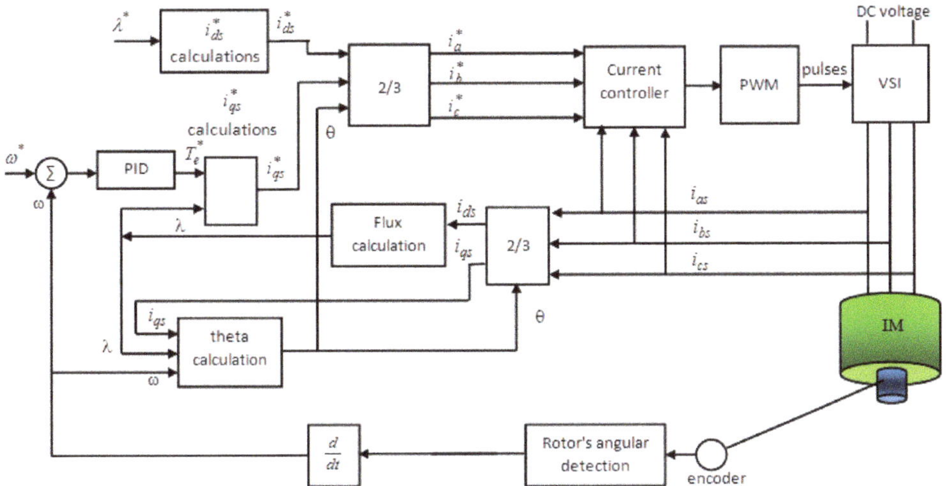


**Figure 8.** Block diagram of the IM controlled via vector control.

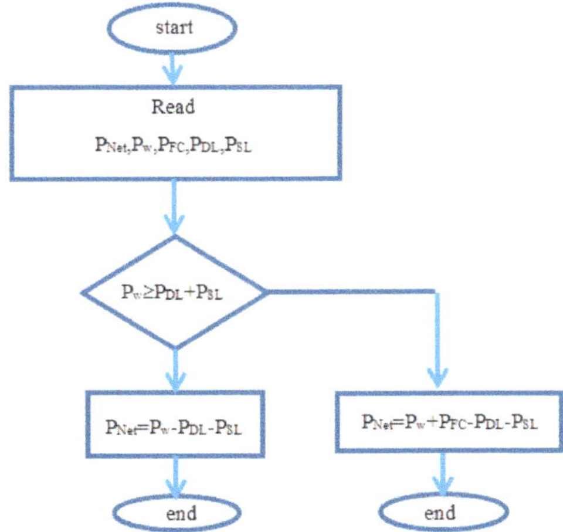


**Figure 9.** The flowchart of the power exchange strategy in the proposed autonomous microgrid.

Per unit DC-link actual and desired load voltage and its enlarging are shown in Figure 10h,i, respectively. These figures show that the applied controller tracks well the desired load voltage. It is clear that the response has little overshoot and settling time against all disturbances. This leads also to constant AC output voltage of the inverter as shown in Figure 10j,k.

The FC power increased at a time of 2 s, where the static load current is increased as shown in Figure 10g,l. Also, Figure 10g shows that the fuel cell generated power is more increased in time 2.25 s, where the wind power is decreased. However, the FC power is increased to recover the load power increase. Figure 10g shows also that the FC power decreased when the static load current decreased at time 4.5 s as shown in Figure 10l,m.

Figure 10n,o show the speed response of the induction motor and its enlarging respectively. From these figures it is seen that the vector controller tracks very well the reference speed of the induction motor. As indicated there is no overshoot and without settling time. Figure 10q shows the stator current of the IM at different speeds and torques. However, the IM load torque changes between 7 and 10 Nm and the IM speed varies between 120 and 200 rad/s, as shown in Figure 10p,n, respectively.

The obtained results are compared with the results obtained in [37], where sliding mode control (SMC) and NARMA-control are applied. The results show that both the proposed optimal control and the robust SMC are able to achieve good voltage and current waveforms parameters and to track the reference DC-link voltage and motor speed with very small overshoot and zero steady state error.

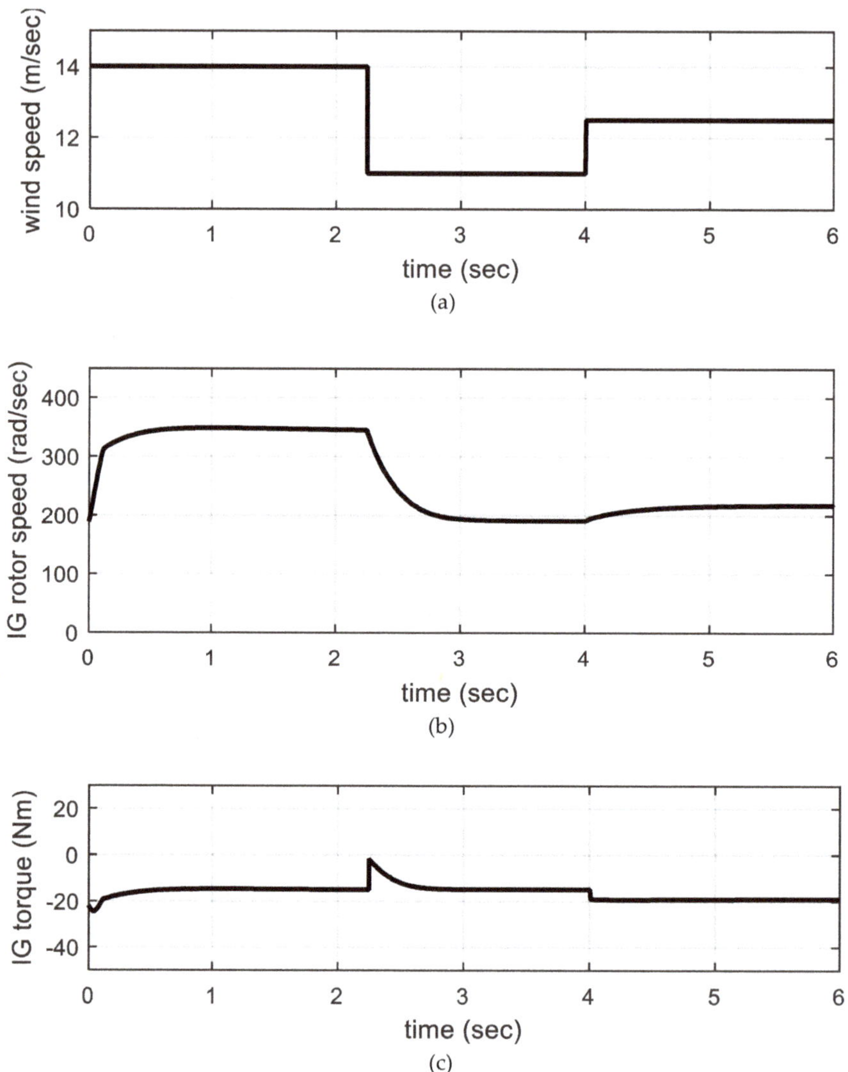


**Figure 10.** *Cont.*

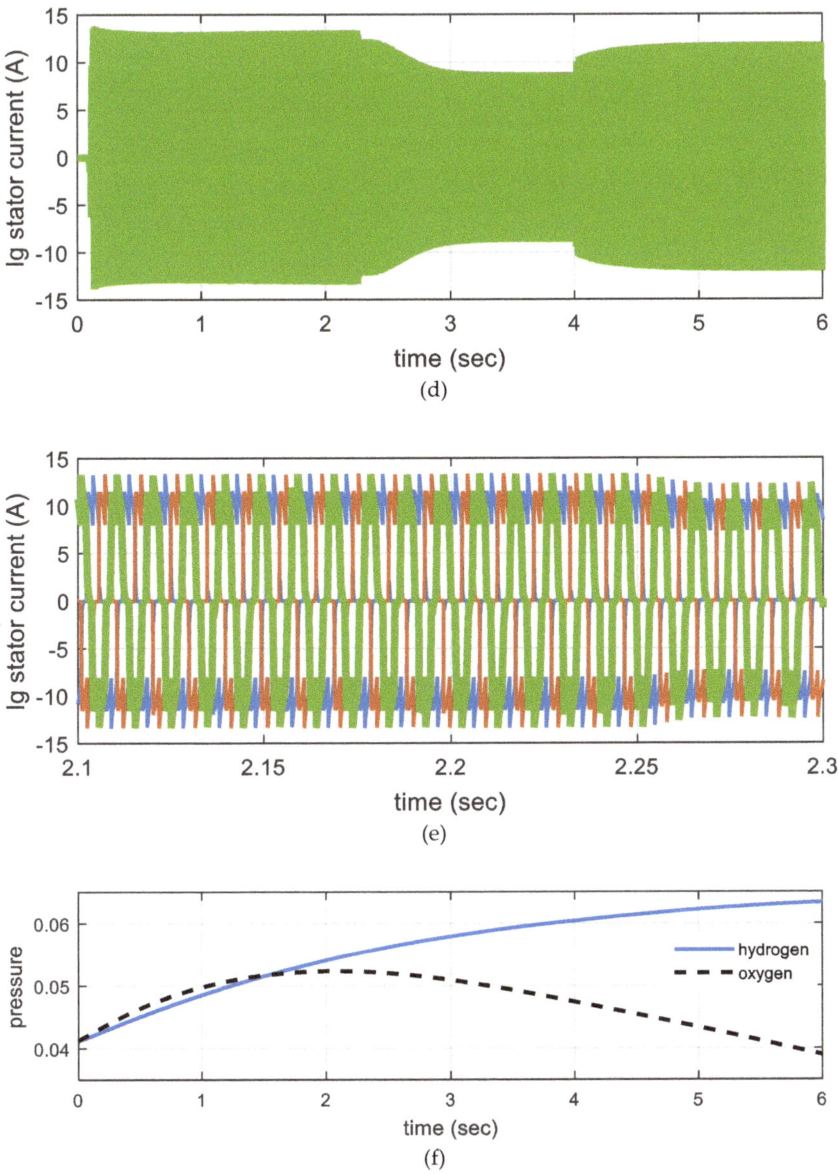


**Figure 10.** *Cont.*

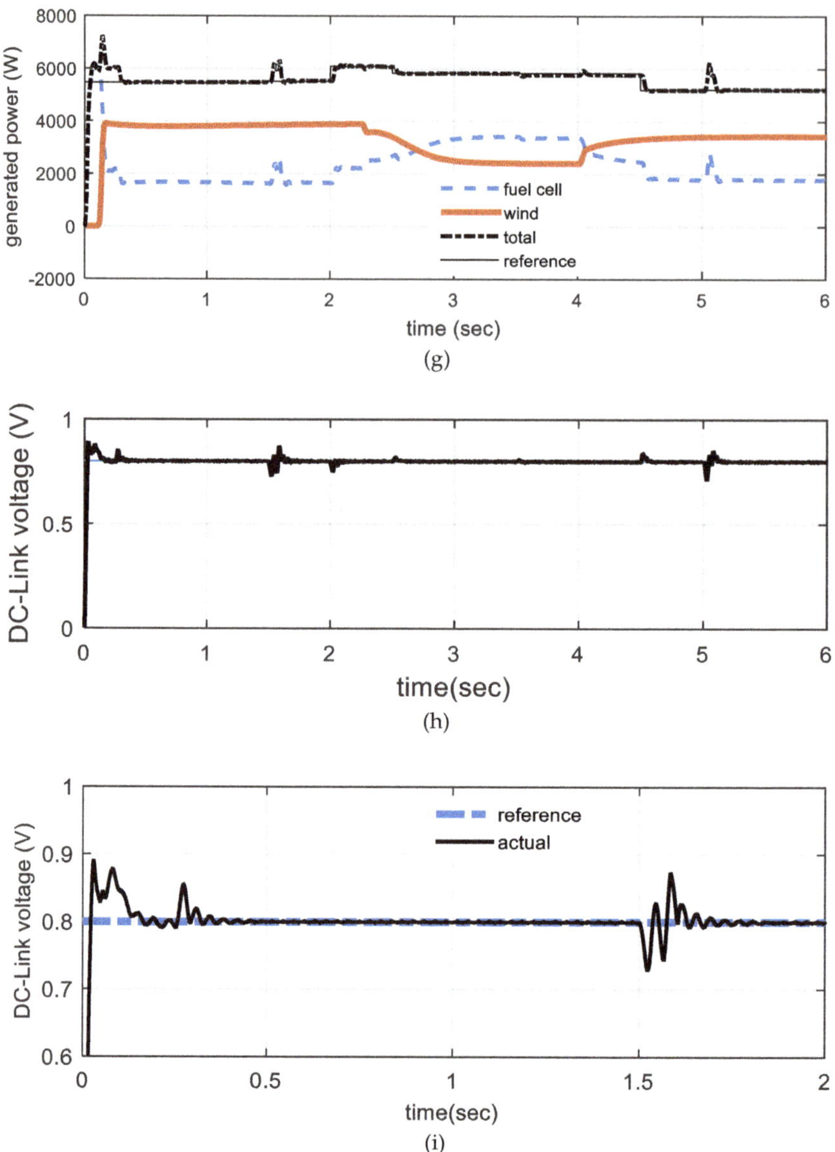


**Figure 10.** *Cont.*

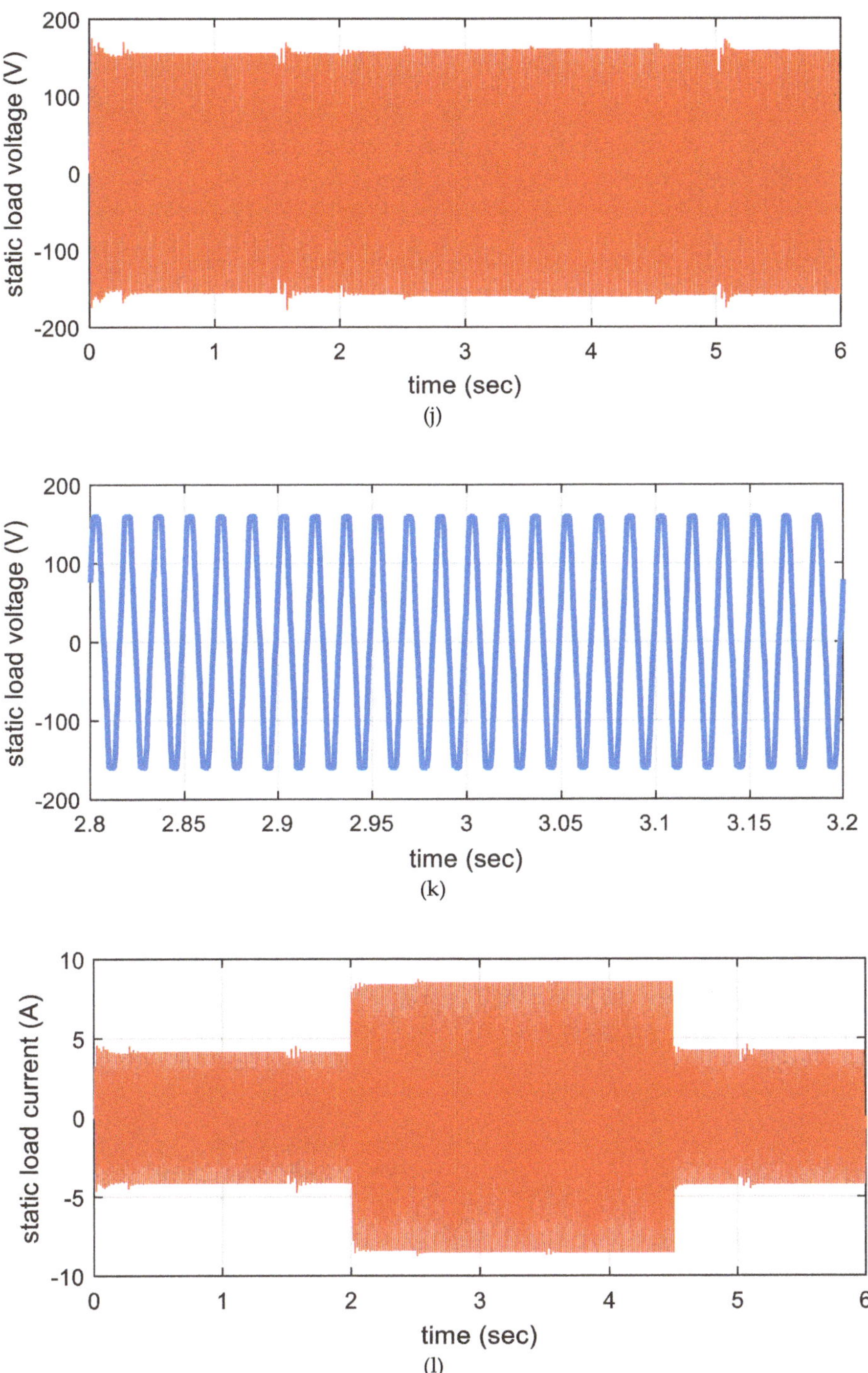


**Figure 10.** *Cont.*

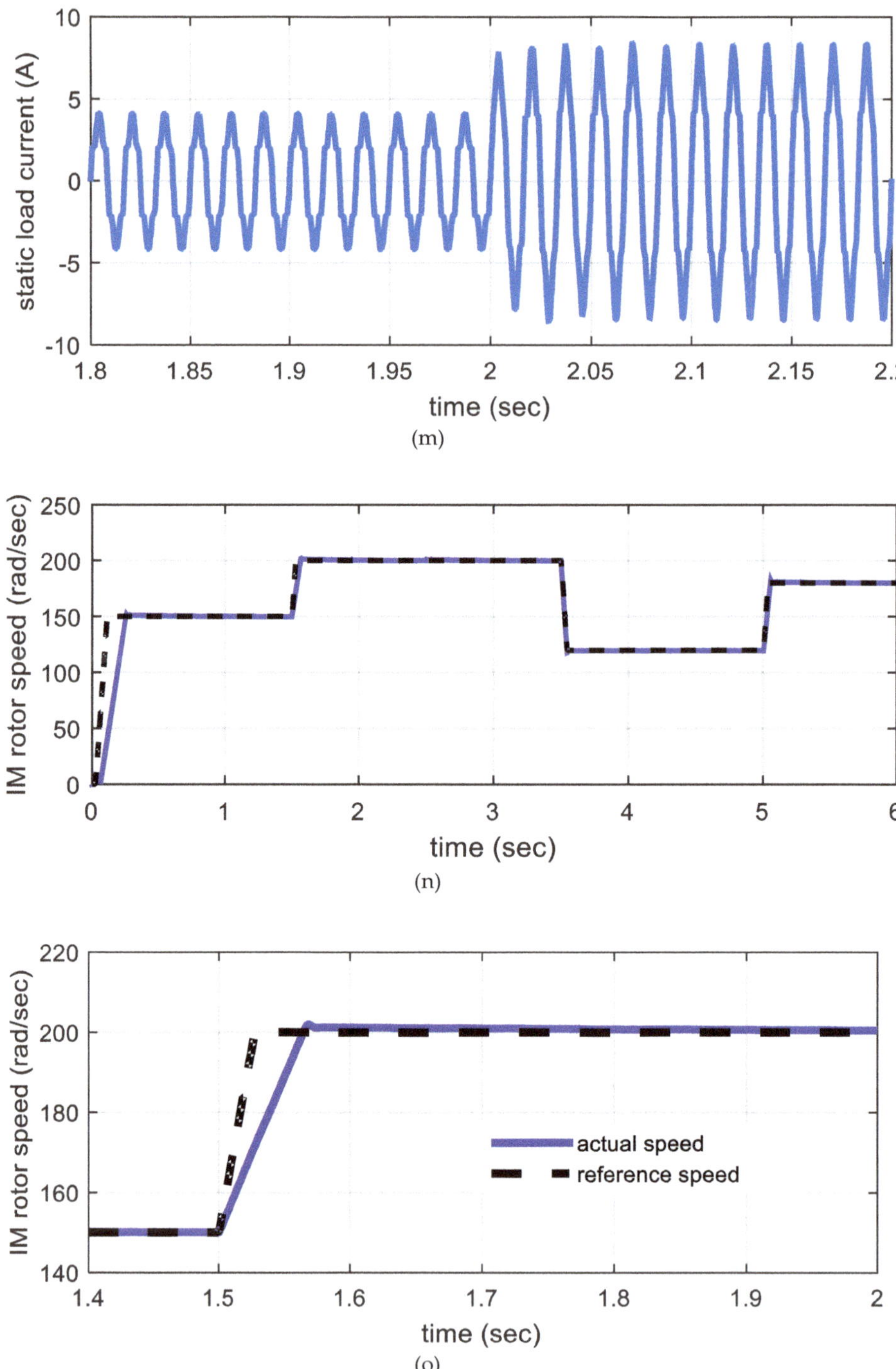


(m)

(n)

(o)

**Figure 10.** *Cont.*

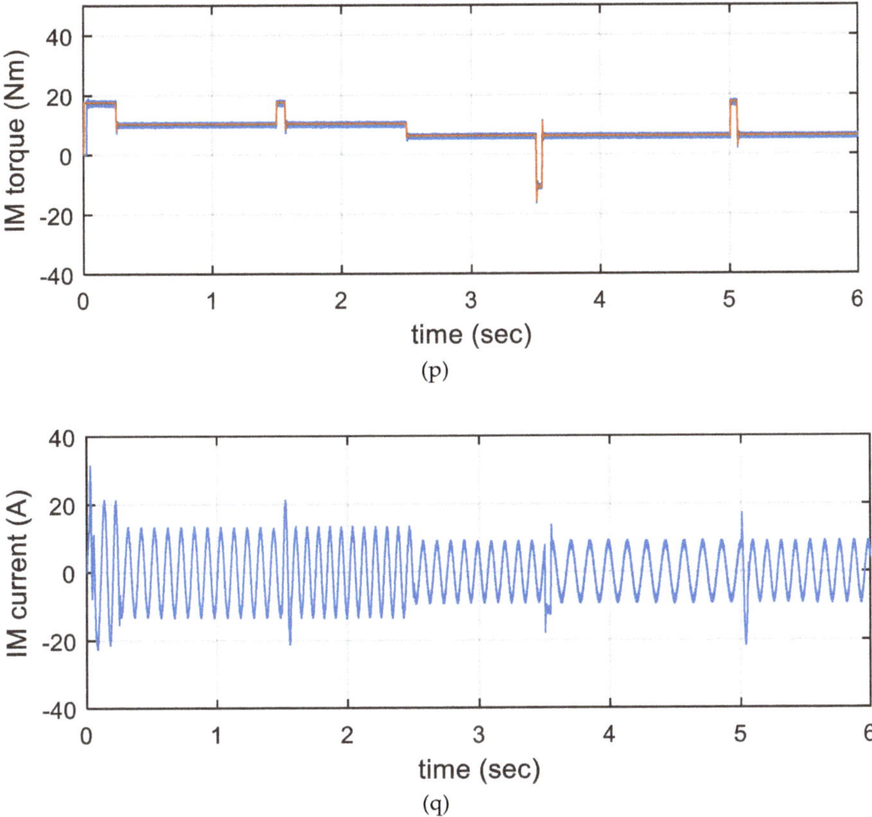


**Figure 10.** Simulation results of the proposed system. (**a**) Wind velocity, (**b**) IG speed, (**c**) IG torque, (**d**) IG stator current, (**e**) enlarging of (**d**), (**f**) FC pressure of $H_2$ and $O_2$, (**g**) generated power, (**h**) DC-link voltage (pu), (**i**) enlarging of (**h**), (**j**) static load voltage, (**k**) enlarging of (**j**), (**l**) static load current, (**m**) enlarging of (**l**), (**n**) IM speed, (**o**) enlarging of (**n**), (**p**) IM torque, and (**q**) IM stator current.

## 6. Conclusions

This article proposed a microelectrical power grid system composed of an optimal controller design using an MBA algorithm. This studied controlled power system mainly includes hybrid wind/fuel cell generation unit which feeding both dynamic and static loads. These loads are fed by the fuel cell and the wind power generation system. At high values of wind speed wind power acts as the master source that supplies the loads and store hydrogen in the FC via the electrolyzer. Consequently, at low values of the wind speed, the FC acts as a slave that supplies the loads.

IM as a dynamic load, a series R-L load, and as a static load are considered in this paper. The main inverter controller has two nested control loops. The outer loop (voltage loop) uses an optimal PI controller, while the inner loop (current loop) uses hysteresis controller current. On the other hand, the rotor speed of the IM is controlled using optimal vector control.

The proposed microgrid system is simulated using Simulink/MATLAB software and is tested in step variations of wind speed, IM rotor speed, IM torque, and static load current. The results of the simulation show that the proposed generation system success to supply the loads perfectly under all disturbances. It is indicated that the performance of the main inverter controller is excellent, as the load power responses have low overshoot accompanied by small settling time. Also, the proposed optimal controller is able to maintain the DC-link voltage and hence the AC load voltage at its reference value for any variations in wind velocity and the current of the static load and/or dynamic load parameters variations. We also found that the speed of the IM follows its desired value without any settling time or any overshoot. The obtained results show that both the generated wind and fuel cell powers are generated so that the wind power feeds the load power demand, while the fuel cell power compensates for any extra needed power.

**Author Contributions:** I.E.A., A.M.K., and S.A.Z. conceived, designed the system model, analyzed the results, and wrote the paper.

**Funding:** This research received no external funding.

**Conflicts of Interest:** The authors declare no conflict of interest.

## Nomenclatures

| | |
|---|---|
| $P_m$ | the power output of the wind turbine |
| $\beta$ | the blade pitch angle (in degrees) |
| $\rho$ | the air density (kg/m$^3$) |
| $v_w$ | the wind speed |
| R | the radius of the turbine blade |
| $C_p$ | the performance coefficient of the turbine |
| $\omega_m$ | the turbine mechanical angular rotor speed |
| $T_m$ | the wind turbine torque |
| $T_e$ | the electrical generator electromagnetic torque (N·m), |
| J | the combined inertia of the generator rotor and the wind turbine (kg·m) |
| B | the mechanical viscous friction (N·m·s/rad) |
| $r_r$ | the rotor resistance |
| $v_{ds}$, $v_{qs}$ | the stator d- and q-axis voltage components |
| $i_{ds}$, $i_{qs}$ | the stator d- and q-axis current components |
| $\lambda_{dr}$, $\lambda_{qr}$ | the rotor d- and q-axis flux linkage components |
| $L_s$, $L_r$, $L_m$ | the stator inductance, rotor, and mutual inductances |
| $\omega_r$ | the motor speed, $\omega_s$ is the motor synchronous speed |
| $T_{em}$ | the motor electromagnetic torque (N·m) |
| $J_m$ | the inertia of the IM rotor (kg·m) |
| $B_m$ | the viscous friction of the coupling (N·m·s/rad) |
| E | the stack output voltage |
| $E_o$ | the cell open circuit voltage at standard pressure |
| N | the number of cells in stack |
| F | the Faraday's constant |
| n | the number of transferred electrons in the electrochemical reaction |
| $R'$ | the universal gas constant |
| T | the operating temperature |
| $P_{H2}$ | the partial pressure of hydrogen |
| $P_{O2}$ | the partial pressure of oxygen |
| $P_{H2Oc}$ | the partial pressure of gas water |
| $P_{std}$ | the standard pressure |
| $V_{drop}$ | the voltage losses |
| i | the output current density |
| $i_n$ | the internal current density related to internal current losses |
| $i_o$ | the exchange current density related to activation losses |
| $i_L$ | the limiting current density related to concentration losses |
| $\alpha$ | the area specific resistance related to resistive losses |
| ($I_g$, $V_g$) | the phase rms current and voltage of the IG |
| ($I_d$, $V_d$) | the average rectifier output current and voltage |
| d | the duty ratio of the switch |
| ($V_{fc}$, $I_{fc}$) | the fuel cell output voltage and current |

| | |
|---|---|
| $\underline{V}_c$ | the output voltage, |
| $\underline{V}_i$, | the inverter voltage |
| $\underline{I}_o$ | the output current |
| $\underline{I}_f$ | the filter current |
| $f_a$, $f_b$, $f_c$ | the phase values, |
| $\underline{F}$ | the space vector of the quantity |
| (L, C) | the filter inductance and capacitance |
| $z_o$ | the first shot point, |
| SB, LB | are the problem upper and lower limits |
| $\gamma$ | the exploration factor |
| $N_s$ | the Shrapnel pieces |
| $\vec{d}_{i-1}$ | the shrapnel distance of the exploded mines |
| F | the fitness function |
| $\vec{z}_e$ | the best location |
| $t_s$ | the simulation time |
| $K_{p1}$, $K_{i1}$, $K_{p2}$, $K_{i2}$ | The PI control parameters to be estimated |
| $\underline{P_W}$ | electrical output power of wind generation system |
| $\underline{P_{FC}}$ | electrical output power of fuel cell generation system |
| $\underline{P_{SL}}$ | electrical power needed by the static load |
| $\underline{P_{DL}}$ | electrical power needed by the dynamic load |
| $\underline{P_{Net}}$ | the difference between the generated and demanded powers |

## References


1. REN21. Renewable energy Policy Network for the 21st Century. Global Status Report. 2018. Available online: http://www.ren21.net/gsr_2018_full_report_en (accessed on 20 November 2018).
2. Meisen, P.; Loiseau, A. *Ocean Energy Technologies for Renewable Energy Generation*; Global Energy Network Institute (GENI): New York, NY, USA, 2009.
3. Pena, R.; Cardenas, R.; Proboste, J.; Clare, J.; Asher, G. Wind–diesel generation using doubly fed induction machines. *IEEE Trans. Energy Convers.* **2008**, *23*, 202–214. [CrossRef]
4. Zhou, W.; Lou, C.; Li, Z.; Lu, L.; Yang, H. Current status of research on optimum sizing of stand-alone hybrid solar-wind power generation systems. *Appl. Energy* **2010**, *87*, 380–389. [CrossRef]
5. Sanyal, S.K. Future of geothermal energy. In Proceedings of the 35th Workshop on Geothermal Reservoir Engineering, Stanford, CA, USA, 1–3 February 2010.
6. Lazard's Levelized Cost of Energy Analysis—Version 10. Available online: https://www.lazard.com/media/438038/levelized-cost-of-energy-v100.pdf (accessed on 23 December 2016).
7. Nehrir, M.H.; Wang, C.; Strunz, K.; Aki, H.; Ramakumar, R.; Bing, J.; Miao, Z.; Salameh, Z. A Review of Hybrid Renewable/Alternative Energy Systems for Electric Power Generation: Configurations, Control, and Applications. *IEEE Trans. Sustain. Energy* **2011**, *2*, 392–403. [CrossRef]
8. Barnard, M. 7 Factors Show Wind Solar 1st Choices. Available online: https://cleantechnica.com/2016/07/11/7-factors-show-wind-solar-1st-choices/ (accessed on 11 July 2016).
9. Zhang, D.; Evangelisti, S.; Lettieri, P.; Papageorgiou, L.G. Economic and environmental scheduling of smart homes with microgrid: DER operation and electrical tasks. *Energy Convers. Manag.* **2016**, *110*, 113–124. [CrossRef]
10. Koohi-Kamali, S.; Rahim, N.A. Coordinated control of smart microgrid during and after islanding operation to prevent under frequency load shedding using energy storage system. *Energy Convers. Manag.* **2016**, *127*, 623–646. [CrossRef]
11. Camblong, H.; Baudoin, S.; Vechiu, I.; Etxeberria, A. Design of a SOFC/GT/SCs hybrid power system to supply a rural isolated microgrid. *Energy Convers. Manag.* **2016**, *117*, 12–20. [CrossRef]
12. Chauhan, A.; Saini, R.P. A review on Integrated Renewable Energy System based power generation for stand-alone applications: Configurations, storage options, sizing methodologies and control. *Renew. Sustain. Energy Rev.* **2014**, *38*, 99–120. [CrossRef]
13. The Energy Foundation Annual Report. Available online: http://www.ef.org/ten-signs-in-2016-that-show-the-clean-energy-economy-is-here-to-stay/ (accessed on 23 December 2016).

14. Onar, O.C.; Uzunoglu, M.; Alam, M.S. Dynamic modeling, design and simulation of a wind/fuel cell/ultra-capacitor-based hybrid power generation system. *J. Power Sources* **2006**, *161*, 707–722. [CrossRef]
15. Kassem, A.M.; Hassan, A.A. Performance improvements of a permanent magnet synchronous machine via functional model predictive control. *J. Control Sci. Eng.* **2012**, *2012*, 7. [CrossRef]
16. Ribeiro, P.F.; Johnson, B.K.; Crow, M.L.; Arsoy, A.; Liu, Y. Energy storage systems for advanced power applications. *Proc. IEEE* **2001**, *89*, 1744–1756. [CrossRef]
17. Rastler, D. *Electricity Energy Storage Technology Options: A White Paper Primer on Applications, Costs, and Benefits*; Technical Report; Electric Power Research Institute (EPRI): Palo Alto, CA, USA, 2010.
18. Giraud, F.; Salameh, Z.M. Steady-state performance of a grid-connected rooftop hybrid wind–photovoltaic power system with battery storage. *IEEE Trans. Energy Convers.* **2001**, *16*, 1–7. [CrossRef]
19. Yang, H.X.; Lu, L.; Zhou, W. A novel optimization sizing model for hybrid solar–wind power generation system. *Solar Energy* **2007**, *81*, 76–84. [CrossRef]
20. Baziar, A.; Kavousi-Fard, A. Considering uncertainty in the optimal energy management of renewable micro-grids including storage devices. *Renew. Energy* **2013**, *59*, 158–166. [CrossRef]
21. Ahmed, M.; Amin, U.; Aftab, S.; Ahmed, Z. Integration of Renewable Energy Resources in Microgrid. *Energy Power Eng.* **2015**, *7*, 12–29. [CrossRef]
22. Delfino, B.; Fornari, F. Modeling and Control of an Integrated Fuel Cell/Wind Turbine System. In Proceedings of the 2003 IEEE Bologna Power Tech Conference, Bologna, Italy, 23–26 June 2003; p. 6. [CrossRef]
23. Fathabadi, H. Novel standalone hybrid solar/wind/fuel cell power generation system for remote areas. *Solar Energy* **2017**, *146*, 30–43. [CrossRef]
24. Vidyanandan, K.V.; Senroy, N. Frequency regulation in microgrid using wind—Fuel cell—Diesel generator. In Proceedings of the 2012 IEEE Power and Energy Society General Meeting, San Diego, CA, USA, 22–26 July 2012; pp. 1–8. [CrossRef]
25. Hussaina, A.; Arifb, S.M.; Aslamc, M. Emerging renewable and sustainable energy technologies: State of the art. *Renew. Sustain. Energy Rev.* **2017**, *71*, 12–28. [CrossRef]
26. Al-falahi, M.D.A.; Jayasinghe, S.D.G.; Enshaei, H. A review on recent size optimization methodologies for standalone solar and wind hybrid renewable energy system. *Energy Convers. Manag.* **2017**, *143*, 252–274. [CrossRef]
27. Zhu, Y.; Tomsovic, K. Development of models for analyzing the load-following performance of micro-turbines and fuel cells. *Electr. Power Syst. Res.* **2002**, *62*, 1–11. [CrossRef]
28. Khan, F.I.; Hawboldt, K.; Iqbal, M.T. Life Cycle Analysis of wind–fuel cell integrated system. *Renew. Energy* **2005**, *30*, 157–177. [CrossRef]
29. Battista, H.D.; Mantz, R.J.; Garelli, F. Power conditioning for a wind/hydrogen energy system. *J. Power Sources* **2006**, *155*, 478–486. [CrossRef]
30. Gorgun, H. Dynamic Modeling of a Proton Exchange Membrane (PEM) Electrolyzer. *Int. J. Hydrogen Energy* **2006**, *31*, 29–38. [CrossRef]
31. Bizon, N. Optimal operation of fuel cell/wind turbine hybrid power system under turbulent wind and variable load. *Appl. Energy* **2018**, *212*, 196–209. [CrossRef]
32. Mendis, N.; Muttaqi, K.M.; Sayeef, S.; Perera, S. Standalone operation of wind turbine-based variable speed generators with maximum power extraction capability. *IEEE Trans. Energy Conv.* **2012**, *27*, 822–834. [CrossRef]
33. MATLAB Math Library User's Guide', by the Math Works. Inc. 2015. Available online: https://fenix.tecnico.ulisboa.pt/downloadFile/845043405443232/sl_using_r2015a.pdf (accessed on 1 March 2015).
34. Chiu, L.Y.; Diong, B.; Gemmen, R.S. An improved small-signal model of the dynamic behavior of pem fuel cells. *IEEE Trans. Ind. Appl.* **2004**, *40*, 970–977. [CrossRef]
35. Belyaev, P.V.; Mischenko, V.S.; Podberezkin, D.A.; Em, R.A. Simulation Modeling of Proton Exchange Membrane Fuel Cells. In Proceedings of the 2016 Dynamics of Systems, Mechanisms and Machines (Dynamics), Omsk, Russia, 15–17 November 2016; pp. 1–5. [CrossRef]
36. SharifiAsl, S.M.; Rowshanzamir, S.; Eikani, M.H. Modelling and simulation of the steady-state and dynamic behavior of a PEM fuel cell. *Energy* **2010**, *35*, 1633–1646. [CrossRef]
37. Kassem, A.M. Modeling and Robust Control Design of a Standalone Wind-Based Energy Storage Generation Unit Powering an Induction Motor Variable-Displacement Pressure Compensated Pump. *IET Renew. Power Gener.* **2016**, *10*, 275–286. [CrossRef]

38. Rashid, M. *Power Electronics Handbook*, 2nd ed.; Elsevier Press: California, CA, USA, 2011.
39. Ali, E.S.; Abd Elazim, S.M. Mine blast algorithm for environmental economic load dispatch with valve loading effect. *Neural Comput. Appl.* **2016**, *30*, 261–270. [CrossRef]
40. Sadollah, A.; Bahreininejad, A.; Eskandar, H.; Hamdi, M. Mine blast algorithm: A new population based algorithm for solving constrained engineering optimization problems. *Appl. Soft Comput.* **2013**, *13*, 2592–2612. [CrossRef]
41. Majumdar, S.; Mandal, K.; Chakraborty, N. Performance study of mine blast algorithm for automatic voltage regulator tuning. In Proceedings of the 2014 Annual IEEE India conference (INDICON), Pune, India, 11–13 December 2014; pp. 1–6. [CrossRef]
42. Trzynadlowski, A.M. *The Field Orientation Principle in Control of Induction Motors*, 1st ed.; Springer: New York, NY, USA, 1994.

*Article*

# Design and Implementation of the Off-Line Robust Model Predictive Control for Solid Oxide Fuel Cells †

**Narissara Chatrattanawet [1], Soorathep Kheawhom [1], Yong-Song Chen [2] and Amornchai Arpornwichanop [1,*]**

1 Center of Excellence in Process and Energy Systems Engineering, Department of Chemical Engineering, Faculty of Engineering, Chulalongkorn University, Bangkok 10330, Thailand; narissara_hl@hotmail.com (N.C.); soorathep.k@chula.ac.th (S.K.)

2 Advanced Institute of Manufacturing with High-Tech Innovations and Department of Mechanical Engineering, National Chung Cheng University, Chiayi 62102, Taiwan; imeysc@ccu.edu.tw

* Correspondence: amornchai.a@chula.ac.th

† This paper is an extended version of paper presented at the 12th Process Systems Engineering (PSE) and 25th European Symposium of Computer Aided Process Engineering (ESCAPE), Copenhagen, Denmark, 31 May–4 June 2015.

Received: 14 September 2019; Accepted: 23 November 2019; Published: 3 December 2019

**Abstract:** An off-line robust linear model predictive control (MPC) using an ellipsoidal invariant set is synthesized based on an uncertain polytopic approach and then implemented to control the temperature and fuel in a direct internal reforming solid oxide fuel cell (SOFC). The state feedback control is derived by minimizing an upper bound on the worst-case performance cost. The simulation results indicate that the synthesized robust MPC algorithm can control and guarantee the stability of the SOFC; although there are uncertainties in some model parameters, it can keep both the temperature and fuel at their setpoints.

**Keywords:** solid oxide fuel cell; robust model predictive control; off-line calculation; control synthesis

## 1. Introduction

A solid oxide fuel cell (SOFC) is a promising fuel cell technology that can be used in co-generation systems for widespread commercial applications [1]. No moving parts, quiet operation, low pollution, and high efficiency are the advantages of fuel cells. Many researchers have discussed the considerable environmental benefits of fuel cell technology [2]. Using SOFC technology also involves the depletion of greenhouse gas emissions when compared with traditional energy generation methods. Moreover, there is interest in the development of the fuel cell technology as a substitute for internal combustion [3]. In general, a SOFC is operated over a wide temperature range, from 873–1273 K, which leads to high energy conversion efficiency, fuel flexibility, and the possibility for combined heat and power systems [4].

The SOFC can use various fuel types, such as methane, methanol, ethanol, and other hydrocarbons, due to its high operating temperature range. Even though high-chain hydrocarbons, such as n-dodecane, can be used as a fuel for the SOFC system, methane is generally considered for SOFC operation, due to its availability, highest hydrogen to carbon ratio in hydrocarbon substances and low cost [5,6]. As the long-chain hydrocarbon fuel contains high carbon and it has a low hydrogen-to-carbon ratio, fuel processing is required for breaking this fuel down into small substances and increasing the hydrogen-to-carbon ratio for the avoidance of a carbon formation in SOFC [7]. In general, methane can be synthesized, as a major product or a by-product, from many chemical processes, or even formation process [8]. In addition, methane in biogas can be directly fed to SOFC under dry conditions; however, there is a risk that is associated with contaminants in biogas when it is introduced to commercial SOFCs while using Ni-YSZ anode [9,10]. Therefore, either commercialized material development or fuel processing with cleaning technologies is required. There are many hydrogen production processes to convert methane into hydrogen-rich gas, such as steam reforming, partial oxidation, and autothermal reforming. However, the methane steam reforming process is perhaps the most well-established technology and it is widely used to produce hydrogen in the conventional SOFC system [11]. Internal reforming process can occur within the fuel cell to directly convert hydrocarbon fuel into hydrogen-rich gas since SOFC is operated at high temperatures [12]. The direct internal reforming (DIR) includes the reforming and water gas shift reaction rates and enthalpies, with these reactions occurring on the surface of the anode. The DIR of methane in an anode of the SOFC can possibly be due to the high temperatures that are present in the SOFC anode and it enables high energy conversion efficiency for the system [13]. However, the complete DIR-SOFC showed poorer performance when compared to the DIR-SOFC with partial external reforming, and thus using the pre-reformer with DIR-SOFC might be a suitable operational option [14]. Nevertheless, the internal reforming reaction that occurred at the anode leads to complicated dynamic behavior. Additionally, the steam reforming process is a highly endothermic reaction [15]. The endothermic cooling effect creates a temperature gradient inside the fuel cell stack. The thermal gradient in the cell stack is significantly managed to minimize, because this gradient results in thermal stresses that leads to cell degradation and failure [16,17]. Consequently, efficient control is needed for preventing thermal cracking and ensuring system stability for this process.

The model of SOFC in cell, stack, and system levels has been proposed, and each type of the SOFC model is employed for different purposes, i.e., design, improvement, control, and optimization [18]. Dynamic modeling is especially beneficial for dynamic system analysis, as well as in control design. SOFC operations are often subjected to transient conditions, and, as a result, the fuel cell dynamics have been increasingly considered in modeling activities [19]. Several published works have concentrated on the dynamic modeling and the control of solid oxide fuel cells [20,21]. Li and Choi [22] studied the control of the power output of an SOFC by applying proportional-integral (PI) controllers to maintaining fuel utilization and voltage as the current of the stack changed. To keep the voltage output under load changes, Chaisantikulwat et al. [19] developed an SOFC dynamic model and a feedback control scheme with a PI controller to control cell voltage by manipulating the concentration of $H_2$. The low-order dynamic model that was derived from the step responses was used for designing feedback control. The results showed that the feedback PI controller was able to maintain a constant SOFC voltage for small step changes in the current load. Furthermore, a dynamic model was used to investigate the dynamics of the SOFC stack and design control strategies [23]. A proportional-integral-derivative (PID) controller was implemented to maintain the outlet fuel temperature and the fuel utilization of a planar anode-supported, direct internal reforming solid oxide fuel cell under intermediate-temperature operation. The feed air was used to maintain the outlet fuel temperature when a disturbance in the current density occurred. A control strategy must be more effective in avoiding oscillatory control action as well as in preventing potentially damaging temperature gradients under a higher magnitude of load changes. Stiller et al. [24] developed a dynamic model for the control of an SOFC and gas turbine hybrid system. The SOFC power, fuel utilization, air flow, and cell temperature were controlled while using a proportional–integral–derivative (PID) type controller. However, the conventional PID controller cannot guarantee stability and performance when large disturbances occur.

Model predictive control (MPC), which is a multivariable control algorithm, computes a controller action while using a process model to predict the processes output trajectory in the future [25]. The implementation of MPC requires the identification of an internal process model. Therefore, applying model-based controllers, such as MPC, is more challenging when compared to PI or PID controllers for which explicit controller equations exist [26]. Zhang et al. [25] developed the nonlinear MPC controller (NMPC) for a planar SOFC while using the moving horizon estimation (MHE) method. The current density and molar flow rates of fuel and air were manipulated variables to control the output power, fuel utilization, and cell temperature. The proposed NMPC controller can drive the SOFC following the desired output trajectory when the power output was changed under constant fuel utilization and temperature. In addition, many real chemical processes involve a high degree of parameter uncertainty. Some studies have focused on the development of a robust MPC to handle nonlinear systems and guarantee system stability, as a traditional MPC algorithm is unable to address plant model uncertainties [27,28]. Kothare et al. [29] synthesized a robust MPC algorithm that explicitly incorporated plant model uncertainties. The state feedback control law was obtained by minimizing the worst-case performance cost. This worst-case scenario was used by the simultaneous design and control methodologies to evaluate the process cost function and constraints that were considered in the process [30]. Manthanwar et al. [31] studied the derivation of the explicit control strategy while using a min–max formulation to safeguard against the worst-case uncertainty problem. This guarantee process feasibility as well as process stability for efficient plant operation. A convex optimization problem with linear matrix inequalities (LMIs) constraints was formulated. Bumroongsri and Kheawhom [32] proposed a robust MPC for uncertain polytopic discrete-time systems. In addition, an ellipsoidal off-line MPC strategy for linear parameter varying (LPV) systems was studied. The smallest ellipsoid that contained the present measured state was determined in each sequence of ellipsoids and the scheduling parameter for LPV was measured on-line. Pannocchia [33] also developed a robust MPC algorithm to stabilize the system that was described while using a linear time-varying (LTV) model. Kouramas et al. [34] focused on the design of an MPC controller to control the cell voltage and cell temperature. The results showed that the controller was able to maintain the SOFC voltage and temperature at the desired values. A comprehensive model of SOFC behavior involves numerous complex phenomena, which include electrochemical reaction and the thermal and mechanical properties of the materials. Thus, the SOFC model involves a great deal of parameter uncertainty; the control design should take the model uncertainty into account. For an on-line synthesis approach, the optimization requirement leads to significant amounts of on-line MPC computational time. When MPC incorporates the model uncertainty, the resulting on-line computation will significantly grow with the number of vertices of the uncertainty set. As a result, an off-line synthesis approach is a focus for generating a robust MPC for an uncertain model. With the off-line approach, the computation of a robust MPC is significantly reduced, with only minor losses in its control performance. Wan and Kothare [35] implemented an off-line LMIs for robust MPC while using an asymptotically stable invariant ellipsoid. An off-line robust output feedback MPC approach can certify the robust stability of the closed-loop system in the presence of constraints and it can stabilize both polytopic uncertain systems and norm bound uncertain systems.

A model-based control system should be designed, while taking uncertain parameters into account, to avoid physical damage and achieve high energy efficiency, as the dynamics of SOFC, especially with a DIR operation, are complicated and its model consists of many key parameters. This study concentrates on control design for the DIR SOFC fed by methane-rich gas. The investigation of transient responses of an SOFC by changing the current density, the air and fuel inlet temperatures, and the air and fuel inlet molar flow rates in terms of velocity, is reported. A MIMO control approach using an off-line robust MPC algorithm with LTV system is implemented to control the SOFC with uncertainty in cell voltage. The paper is organized, as follows: Section 2 presents the mathematical model for the SOFC; Section 3 gives a brief review of the robust MPC algorithm; Sections 4 and 5 outline the application, and results and discussion of the proposed off-line robust MPC to the SOFC; and lastly, Section 6 presents the conclusions.

## 2. SOFC Model

The mathematical model of the solid oxide fuel cell (SOFC) consists of mass balances, energy balances, and the electrochemical model. The assumptions for the lumped model of the SOFC are as follows: (i) heat loss to the surroundings is negligible; (ii) all the gases are ideal gas; (iii) pressure gradients inside gas channels are negligible; (iv) the heat capacity of all gases is temperature independent; and, (v) the exit fuel and air temperatures and the cell temperature are the same.

Typically, the SOFC consists of a ceramic ion-conducting electrolyte and two porous electrodes with a sandwich structure (Figure 1). To generate electricity with an SOFC, methane-rich gas is directly fed into the anode side while air, which is the oxidant, is continuously delivered into the cathode side. At the cathode side, oxygen is reduced, which forms oxygen ions. The oxygen ions can diffuse through the ion-conducting electrolyte to the anode/electrolyte interface. At the anode side, the oxygen ions chemically react with hydrogen in the fuel, producing water and electrons. The electrons are transported via the external circuit and back to the cathode/electrolyte interface, thus producing electrical energy. Exhaust gases and heat are also produced by the SOFC as by-products. The reforming, electrochemical, and energetic models are simultaneously solved to obtain an exact solution [36].

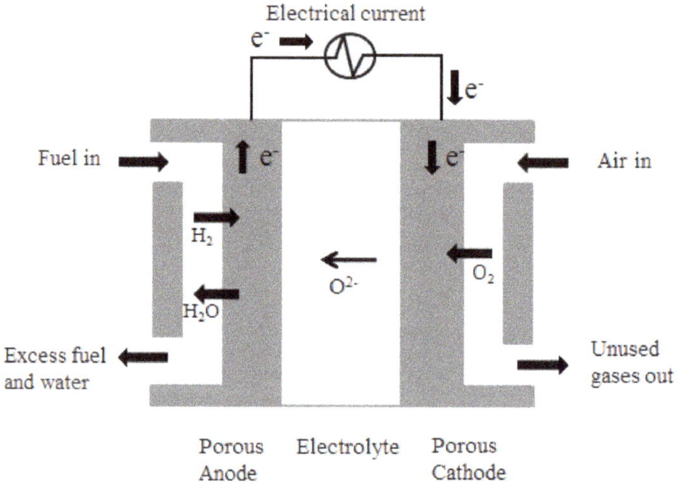


**Figure 1.** Schematic diagram of solid oxide fuel cell operation.

## 2.1. Mass and Energy Balances

The internal methane steam reforming (MSR) reaction in porous-supported SOFC is the most important factor for determining the performance of the SOFC [37]. Furthermore, it is the main reaction for hydrogen production. Table 1 shows the reactions that occurred within the SOFC, which include steam reforming, water-gas shift (WGS), and overall redox reaction [38]. These reactions are used in mass and energy balances. In the fuel channel, it is assumed that methane can only be reformed to hydrogen, carbon monoxide, and carbon dioxide and, therefore, cannot be electrochemically oxidized [23]. In the endothermic steam reforming reaction, fuel in the presence of a catalyst produces hydrogen and carbon monoxide. Table 1 shows the rate expression of the steam reforming reaction $R_{(i)}$ [38], where $k_0$ is the pre-exponential constant, being equal to 4272 mol s$^{-1}$ m$^{-2}$ bar$^{-1}$, and $E_a$ is the activation energy, equal to 82 kJ mol$^{-1}$. Excess steam is used to prevent carbon formation on the catalyst and force the reaction to completion. An associated reaction to the reforming reaction is the water-gas shift reaction. Unlike the steam reforming reaction, the water gas-shift reaction is an exothermic reaction. The rate expression of the water gas shift reaction is written as $R_{(ii)}$. The overall redox reaction $R_{(v)}$ associates with the electric current density ($j$), according to Faraday's law.

**Table 1.** Reactions and reaction rates considered in a solid oxide fuel cell (SOFC) [38].

| Reactions | No. | Reaction Equations | Reaction Rates | ΔH (kJ mol$^{-1}$) |
|---|---|---|---|---|
| Steam reforming | (i) | $CH_4 + H_2O \leftrightarrow 3H_2 + CO$ | $R_{(i)} = k_0 p_{CH_4} \exp\left(-\frac{E_a}{RT}\right)$ | 206.10 |
| Water-gas shift | (ii) | $CO + H_2O \leftrightarrow H_2 + CO_2$ | $R_{(ii)} = k_{WGSR} p_{CO}\left(1 - \frac{p_{CO_2} p_{H_2} / p_{CO} p_{H_2O}}{K_{eq}}\right)$ | −41.15 |
| Hydrogen oxidation | (iii) | $H_2 + O^{2-} \rightarrow H_2O + 2e^-$ | | |
| Oxygen reduction | (iv) | $1/2O_2 + 2e^- \rightarrow O^{2-}$ | | |
| Overall redox reaction | (v) | $H_2 + 1/2O_2 \rightarrow H_2O$ | $R_{(v)} = \frac{j}{2F}$ | −241.83 |

The lumped-parameter modeling, when only considering changes in time, is a simple approach for describing the dynamic modeling of the solid oxide fuel cell. Xi et al. [39] showed that lumped-parameter models are adequate for systems-level analysis and control through experimental validation. Moreover, the lumped model has been implemented for analysis and control of the planar SOFC systems [40]. Consequently, this work has used the lumped-parameter model for analysis, design, and control of the SOFC.

Equations (1) and (2), respectively, give the mass balances in the fuel and air channels, which provide the amount in moles of each species in the SOFC. The gas compositions in the fuel channel consist of $CH_4$, $H_2O$, $CO$, $H_2$, and $CO_2$, while $O_2$ and $N_2$ are the gas species in the air channel. The mass balances are:

$$\frac{dn_{i,f}}{dt} = \dot{n}_{i,f}^{in} - \dot{n}_{i,f} + \sum_{k\in\{(i),(ii),(v)\}} v_{i,k} R_k A \quad (1)$$

$$\frac{dn_{i,a}}{dt} = \dot{n}_{i,a}^{in} - \dot{n}_{i,a} + v_{i,(v)} R_{(v)} A \quad (2)$$

where $\dot{n}_{i,f}$ and $\dot{n}_{i,a}$ are the molar flow rate of species $i$ in the fuel and air channels, respectively; $v_{i,k}$ is the stoichiometric coefficient of component $i$ in reaction $k$; $R_k$ is the rate of reaction k; and, $A$ is a reaction area.

The temperature change within the cell is neglected for the energy balance. Equations (3) and (4) are used to compute the SOFC temperature ($T_{FC}$).

$$\frac{dT_{FC}}{dt} = \frac{1}{\rho_{SOFC} Cp_{SOFC} V_{SOFC}} \left( \dot{Q}_{f,in} - \dot{Q}_{f,out} + \dot{Q}_{a,in} - \dot{Q}_{a,out} + \sum_{k\in\{(i),(ii),(v)\}} (-\Delta H)_k R_k A - jAV_{FC} \right) \quad (3)$$

$$\dot{Q}_i = \sum_j \dot{n}_j Cp_j (T_i - T_{ref}) \quad (4)$$

where $\dot{Q}_i$ is the enthalpy flow in/out each fuel cell channel.

### 2.2. Electrochemical Model

The Nernst equation explained the difference between the thermodynamic potentials of the electrode reactions is used (Equation (5)) to determine the reversible cell voltage or theoretical open-circuit voltage ($E_{OCV}$).

$$E_{OCV} = E_0 - \frac{RT_{FC}}{2F} \ln\left( \frac{p_{H_2O}}{p_{H_2} p_{O_2}^{0.5}} \right) \quad (5)$$

where $E_0$ is the open-circuit potential at the standard pressure, which is related to the SOFC temperature, as shown in Equation (6) [41].

$$E_0 = 1.253 - 2.4516 \times 10^{-4} T_{FC}(\mathrm{K}) \tag{6}$$

When an external load is combined, the actual voltage ($V_{FC}$) is lower than the open-circuit voltage, owing to the voltage losses: ohmic losses ($\eta_{\mathrm{Ohm}}$), concentration overpotentials ($\eta_{\mathrm{conc}}$), and activation overpotentials ($\eta_{\mathrm{act}}$), which rely on the SOFC temperature, current density, and fuel compositions. Consequently, the cell voltage can be calculated by subtracting the open-circuit voltage with the voltage drops due to the various losses from the theoretical open circuit voltage, as reported by Aguiar et al. [23]:

$$V_{FC} = E_{OCV} - (\eta_{\mathrm{Ohm}} + \eta_{\mathrm{conc}} + \eta_{\mathrm{act}}) \tag{7}$$

$$\eta_{\mathrm{Ohm}} = j\left(\frac{\tau_{\mathrm{anode}}}{\sigma_{\mathrm{anode}}} + \frac{\tau_{\mathrm{electrolyte}}}{\sigma_{\mathrm{electrolyte}}} + \frac{\tau_{\mathrm{cathode}}}{\sigma_{\mathrm{cathode}}}\right) \tag{8}$$

$$\eta_{\mathrm{conc}} = \frac{RT_{FC}}{2F}\ln\left(\frac{p_{\mathrm{H_2O,TPB}}\,p_{\mathrm{H_2}}}{p_{\mathrm{H_2O}}\,p_{\mathrm{H_2,TPB}}}\right) + \frac{RT_{FC}}{4F}\ln\left(\frac{p_{\mathrm{O_2}}}{p_{\mathrm{O_2,TPB}}}\right) \tag{9}$$

$$p_{\mathrm{H_2,TPB}} = p_{\mathrm{H_2,f}} - \frac{RT\tau_{\mathrm{anode}}}{2FD_{\mathrm{eff,anode}}}j \tag{10}$$

$$p_{\mathrm{H_2O,TPB}} = p_{\mathrm{H_2O,f}} + \frac{RT\tau_{\mathrm{anode}}}{2FD_{\mathrm{eff,anode}}}j \tag{11}$$

$$p_{\mathrm{O_2,TPB}} = P - (P - p_{\mathrm{O_2}})\exp\left(\frac{RT\tau_{\mathrm{cathode}}}{4FD_{\mathrm{eff,cathode}}P}j\right) \tag{12}$$

$$j = j_{0,\mathrm{anode}}\left[\frac{p_{\mathrm{H_2,TPB}}}{p_{\mathrm{H_2}}}\exp\left(\frac{\alpha nF}{RT_{FC}}\eta_{\mathrm{act,anode}}\right) - \frac{p_{\mathrm{H_2O,TPB}}}{p_{\mathrm{H_2O}}}\exp\left(-\frac{(1-\alpha)nF}{RT_{FC}}\eta_{\mathrm{act,anode}}\right)\right] \tag{13}$$

$$j = j_{0,\mathrm{anode}}\left[\frac{p_{\mathrm{H_2,TPB}}}{p_{\mathrm{H_2}}}\exp\left(\frac{\alpha nF}{RT_{FC}}\eta_{\mathrm{act,anode}}\right) - \frac{p_{\mathrm{H_2O,TPB}}}{p_{\mathrm{H_2O}}}\exp\left(-\frac{(1-\alpha)nF}{RT_{FC}}\eta_{\mathrm{act,anode}}\right)\right] \tag{14}$$

$$j_{0,\{\mathrm{anode,\ cathode}\}} = \frac{RT_{FC}}{nF}k_{\{\mathrm{anode,\ cathode}\}}\exp\left(-\frac{E_{\{\mathrm{anode,\ cathode}\}}}{RT_{FC}}\right) \tag{15}$$

where $\tau_{\mathrm{anode}}$, $\tau_{\mathrm{electrolyte}}$, and $\tau_{\mathrm{cathode}}$ are the thickness of the anode, electrolyte, and cathode layers, respectively; $\sigma_{\mathrm{anode}}$ and $\sigma_{\mathrm{cathode}}$ are the electronic conductivity of the anode and cathode, respectively; $\sigma_{\mathrm{electrolyte}}$ is the ionic conductivity of the electrolyte; $p_{i,\mathrm{TPB}}$ is the partial pressure of component $i$ at three-phase boundaries (TPB); $D_{\mathrm{eff,anode}}$ stands for the effective diffusivity coefficient in the anode, while considering a binary gas mixture of $H_2$ and $H_2O$ with equi-molar, counter-current, one-dimensional diffusion due to a major difference in the concentration of these two key components at TPB and flow channel that are caused by the electrochemical reaction [23]; $D_{\mathrm{eff,cathode}}$ stands for the oxygen effective diffusivity coefficient in the cathode (a binary gas mixture of $O_2$ and $N_2$) (the diffusion coefficient for the electrode is assumed to be constant [42]); $\alpha$ is the fraction of the applied potential that promotes the transfer coefficient, which is usually taken to be 0.5 [23]; and, $n$ is the number of electrons that are transferred in the single elementary rate-limiting reaction step represented by the Butler–Volmer equation. The activation energies of the electrode exchange current densities ($E_{\{\mathrm{anode,\ cathode}\}}$) are 137 and 140 kJ mol$^{-1}$ for the cathode and anode, respectively [23]. The pre-exponential factors of the cathode and anode exchange current densities ($k_{\{\mathrm{anode,\ cathode}\}}$) are $2.35 \times 10^{11}$ and $6.54 \times 10^{11}$ $\Omega^{-1}$ m$^{-1}$, respectively [23].

The power density ($P$) is the amount of power per unit area, which can be determined by multiplying the cell voltage by the current density, as expressed:

$$P = jV_{FC} \tag{16}$$

The fuel utilization factor ($U_{\text{fuel}}$) is the ratio between the total fuel consumption for electricity production and the total inlet fuel, as defined:

$$U_{\text{fuel}} = \frac{\bar{j}LW}{(8Fy^0_{\text{CH}_4} + 2Fy^0_{\text{H}_2} + 2Fy^0_{\text{CO}})F^0_{\text{fuel}}} \tag{17}$$

The air ratio ($\lambda_{\text{air}}$) is the inverse of the air utilization factor, which is defined as:

$$\lambda_{\text{air}} = \frac{y^0_{\text{O}_2}F^0_{\text{air}}}{\bar{j}LW/4F} \tag{18}$$

where $L$ is the cell length (cm$^2$), $W$ is the cell width (cm$^2$), $\bar{j}$ is the average current density (A cm$^{-2}$), $F$ is the Faraday constant (C mol$^{-1}$), $y^0_i$ is the mole fraction (–), and $F^0_i$ is the molar flow rate (mol s$^{-1}$).

## 3. Robust Model Predictive Control

A linear time-varying (LTV) system is defined for a multi-model paradigm or polytopic uncertainty to synthesize a robust controller:

$$\begin{aligned} \mathbf{x}(k+1) &= \mathbf{A}(k)\mathbf{x}(k) + \mathbf{B}(k)\mathbf{u}(k) \\ \mathbf{y}(k) &= \mathbf{C}\mathbf{x}(k) \\ \begin{bmatrix} \mathbf{A}(k) & \mathbf{B}(k) \end{bmatrix} &\in \Omega \end{aligned} \tag{19}$$

where $\mathbf{x}(k)$ is the state of the plant, $\mathbf{u}(k)$ is the control input, and $\mathbf{y}(k)$ is the plant output. Furthermore, the set $\Omega$ to be the polytope for polytopic systems is defined as:

$$\Omega = \text{Co}\left\{\begin{bmatrix} \mathbf{A}_1 & \mathbf{B}_1 \end{bmatrix}, \begin{bmatrix} \mathbf{A}_2 & \mathbf{B}_2 \end{bmatrix}, \ldots, \begin{bmatrix} \mathbf{A}_L & \mathbf{B}_L \end{bmatrix}\right\} \tag{20}$$

where Co represents the convex hull and $\begin{bmatrix} \mathbf{A}_i & \mathbf{B}_i \end{bmatrix}$ are the vertices in the convex hull. If the system is the nominal linear time-invariant (LTI) model, it follows that $L = 1$. For other cases, $\begin{bmatrix} \mathbf{A}(k) & \mathbf{B}(k) \end{bmatrix} \in \Omega$, being defined by $L$ vertices as:

$$\begin{aligned} \begin{bmatrix} \mathbf{A}(k) & \mathbf{B}(k) \end{bmatrix} &= \sum_{i=1}^{L} \lambda_i \begin{bmatrix} \mathbf{A}_i & \mathbf{B}_i \end{bmatrix} \\ \sum_{i=1}^{L} \lambda_i = 1, &\quad 0 \le \lambda_i \le 1 \end{aligned} \tag{21}$$

The nonlinear system can be represented by a polytopic uncertain linear time-varying system. Liu [43] has shown that every trajectory ($\mathbf{x}$, $\mathbf{u}$) of a nonlinear system is a trajectory of Equation (19) for some linear time-varying system in the polytope ($\Omega$).

### 3.1. Robust MPC Algorithm

In this section, the explanation of the robust constrained MPC problem that is constituted of input and output constraints integrated with linear matrix inequality (LMI) constraints is presented. At each sampling time $k$, a robust performance objective is a min–max problem (minimization of worst-case performance cost) in terms of the quadratic objective for the LTV system, which is given by Equation (22):

$$\begin{aligned} &\min_{\mathbf{u}(k+i|k), i=0,1,\ldots,m} \quad \max_{\begin{bmatrix} \mathbf{A}(k+i) & \mathbf{B}(k+i) \end{bmatrix} \in \Omega, i \ge 0} \quad J_\infty(k) \\ &J_\infty(k) = \sum_{i=0}^{\infty} \left[\mathbf{x}(k+i|k)^T \mathbf{Q}_1 \mathbf{x}(k+i|k) + \mathbf{u}(k+i|k)^T \mathbf{R} \mathbf{u}(k+i|k)\right] \end{aligned} \tag{22}$$

where $\mathbf{Q}_1 > 0$ and $\mathbf{R} > 0$ are the symmetric weighting matrices.

The optimization problem at each sample time step is formulated as a convex optimization problem that is related to linear matrix inequalities constraints [29]. The Lyapunov function $V(i,k)$, which is defined as: $V(i,k) = \mathbf{x}(k + i/k)^T\mathbf{P}(i,k)\mathbf{x}(k + i/k)$, where $\forall k, \forall i \geq 0$ and $\mathbf{P}(i,k) > 0$, is utilized to ensure stability for the MPC algorithm. It is noted that, for a vector $\mathbf{x}$, $\mathbf{x}(k/k)$ represents the state measured at real time $k$, and $\mathbf{x}(k + i/k)$ represents the state at prediction time $k + i$ predicted at real time $k$.

### 3.2. Off-Line Robust MPC Algorithm Using Ellipsoidal Invariant Sets

The state-feedback control law can be defined as:

$$\mathbf{u}(k+i|k) = F\mathbf{x}(k+i|k), \quad i \geq 0 \tag{23}$$

The state feedback gains $F$ in the control law are defined as $F = Y_iQ_i^{-1}$ to stabilize the closed-loop system within the ellipsoidal invariant set $\varepsilon = \left\{\mathbf{x}|\mathbf{x}^T\mathbf{Q}^{-1}\mathbf{x} \leq 1\right\}$. The matrix variables $Q_i > 0$ and $Y_i$ are achieved from the result of the linear objective minimization problem $J_\infty(k)$, with the upper bound $\gamma$ on the worst-case MPC. The symbol $*$ represents the corresponding transpose of the lower block part of the symmetric matrices. Therefore, it is determined that:

$$\min_{\gamma, Q_i, Y_i} \gamma \tag{24}$$

subject to

$$\begin{bmatrix} 1 & * \\ x_i & Q_i \end{bmatrix} \geq 0 \tag{25}$$

$$\begin{bmatrix} Q_i & * & * & * \\ A_jQ_i + B_jY_i & Q_i & * & * \\ Q_1^{1/2}Q_i & 0 & \gamma I & * \\ R^{1/2}Y_i & 0 & 0 & \gamma I \end{bmatrix} \geq 0, \quad j = 1, 2, \ldots, L \tag{26}$$

Input constraints that are limited by the process equipment impose hard constraints on the manipulated variable $\mathbf{u}(k)$. Boyd et al. [44] proposed the basic idea to handle these constraints for continuous-time systems. However, the discrete-time robust MPC is presented here, as follows:

$$\begin{bmatrix} X & * \\ Y_i^T & Q_i \end{bmatrix} \geq 0, \tag{27}$$

with

$$X_{hh} \leq u^2_{h,\max}, \quad h = 1, 2, \ldots, n_u \tag{28}$$

For output constraints, performance terms impose constraints on the process output y($k$), as:

$$\begin{bmatrix} S & * \\ \left(A_jQ_i + B_jY_i\right)^T C^T & Q_i \end{bmatrix} \geq 0, \tag{29}$$

with

$$S_{rr} \leq y^2_{r,\max}, \quad r = 1, 2, \ldots, n_y \tag{30}$$

It is noted that Equation (27) is used to guarantee input constraint satisfaction, whereas Equation (29) is used to guarantee output constraint satisfaction.

## 4. SOFC Operation

In this work, the SOFC models that are mentioned above are implemented and simulated while using Matlab for analysis, design, and controls study of the SOFC. The lumped parameter model of SOFC is created by the relation between mass and energy balances and is used to investigate steady state and dynamic behavior. Table 2 shows the model parameters and operating conditions for the

SOFC. Regarding the steady-state analysis of the cell voltage, power density, and cell temperature related to the current density, the SOFC is designed to be operated at a current density ($j$) of 0.45 A $cm^{-2}$, at which the SOFC efficiency is optimized [45]. Under this operating condition, the cell voltage ($V_{FC}$) is 0.72 V, the power density ($P$) is 0.32 W $cm^{-2}$, and the cell temperature ($T_{FC}$) is 1058 K.

**Table 2.** Model parameters and operating conditions used in SOFC simulation.

| | | |
|---|---|---|
| ***Model Parameters*** | | |
| Anode effective diffusivity coefficient ($D_{eff,anode}$) | $3.66 \times 10^{-5}$ | $m^2\ s^{-1}$ |
| Cathode effective diffusivity coefficient ($D_{eff,cathode}$) | $1.37 \times 10^{-5}$ | $m^2\ s^{-1}$ |
| Electronic conductivity of anode ($\sigma_{anode}$) | $9.5 \times 10^7/T_{FC}\ \exp(-1150/T_{FC})$ | $\Omega^{-1}\ m^{-2}$ |
| Electronic conductivity of cathode ($\sigma_{cathode}$) | $4.2 \times 10^7/T_{FC}\ \exp(-1200/T_{FC})$ | $\Omega^{-1}\ m^{-2}$ |
| Ionic conductivity of electrolyte ($\sigma_{electrolyte}$) | $33.4 \times 10^3\ \exp(-10{,}300/T_{FC})$ | $\Omega^{-1}\ m^{-2}$ |
| Anode thickness ($\tau_{anode}$) | 500 | μm |
| Cathode thickness ($\tau_{cathode}$) | 50 | μm |
| Electrolyte thickness ($\tau_{electrolyte}$) | 20 | μm |
| Fuel channel height ($h_f$) | 1 | mm |
| Air channel height ($h_a$) | 1 | mm |
| Cell length ($L$) | 0.4 | m |
| Cell width ($W$) | 0.1 | m |
| ***Operating conditions*** | | |
| Pressure ($P$) | 1 | bar |
| Fuel inlet temperature ($T_f^0$) | 1023 | K |
| Air inlet temperature ($T_a^0$) | 1023 | K |
| Fuel utilization factor ($U_{fuel}$) | 70% | |
| Air ratio ($\lambda_{air}$) | 8.5 | |
| Fuel feed | S/C = 2, 10% pre-reforming | |
| Air feed | 21% $O_2$, 79% $N_2$ | |

## 5. Results and Discussion

### 5.1. Dynamics of SOFC

In this part, the dynamic behavior and performance of the SOFC simulated by using nonlinear mass and energy balance equations (Section 2), coupled with initial operating conditions (Section 4), are given. By varying the current density, the inlet air and fuel temperatures, and the inlet air and fuel molar flow rates, the responses of the cell temperature and cell voltage are investigated. Figure 2 shows the open-loop response of the cell temperature and voltage that result from step changes of ±10% in the current density, the inlet air and fuel temperatures, and the inlet air and fuel molar flow rates, given in terms of velocity. It can be seen that, in the initial period between 0 and 3000 s, the cell voltage and cell temperature move to 0.72 V and 1058 K, respectively, which are the nominal operating points in the dynamic model. Step changes of +10% in each input are present during the second period. For the third period, which occurs between 6000 and 9000 s, the responses are a result of step changes of −10% in each input. The last period shows a return to the initial conditions.

As seen in Figure 2a, cell voltage depends on the cell operating temperature, which relies on the gas inlet temperatures and current density. The transient response of the cell voltage and temperature is considered by a step change in current density, with the gas inlet temperatures and molar flow rates maintaining the nominal values. The result provides that the increase of cell operating temperature is caused by an instantaneous increase in hydrogen consumption within the cell when the SOFC is operated at high current densities. In addition, the cell voltage suddenly drops, which is associating with ohmic losses, although the increase in cell temperature will ultimately decrease the activation overpotentials and the internal resistance in ohmic losses. The cell voltage is dependent on the magnitude of ohmic losses [40]. The step changes of ±10% in the inlet air and fuel temperatures are also investigated with the other inputs being kept at their nominal values (Figure 2b,c). The results show that the dynamic response of fuel cell voltage and cell temperature depend on the inlet temperatures of the fuel and air. The increase in fuel and air inlet temperatures causes an increase in the cell temperature and voltage. However, it can be seen that the air inlet temperature significantly affects the fuel cell voltage and temperature. High air feed increases the heat input to the fuel cell, which promotes the reforming reaction rate; more $H_2$ generated leads to high power generation. The inlet flow rates of both air and fuel can be expressed in terms of air and fuel velocity, being calculated from air ratio and fuel utilization factor, respectively. Figure 2d,e show the transient responses of fuel cell voltage and cell temperature for step changes of ±10% in the inlet flow rate of air and fuel. All other inputs, such as the current density and the inlet temperatures of air and fuel, are kept constant. Changes in the fuel cell voltage and cell temperature are observed when the molar flow rates are changed. The increase in inlet molar flow rate of fuel causes an increase in the hydrogen production rate, which is associated with an increase in the reforming reaction rate. Heat for the reformer, which is provided by heat produced in the fuel cell, results in a decrease of the cell temperature, because of the endothermic steam reforming reaction. However, the fuel cell voltage increases due to an increase in the partial pressure of hydrogen and the partial pressure of water decreases as a result of an increase in the fuel flow rate. It is noted that the overshoots in fuel cell voltage that occurred after step changes could be attributed to numerical errors, which are generated by the discontinuity in time [19].

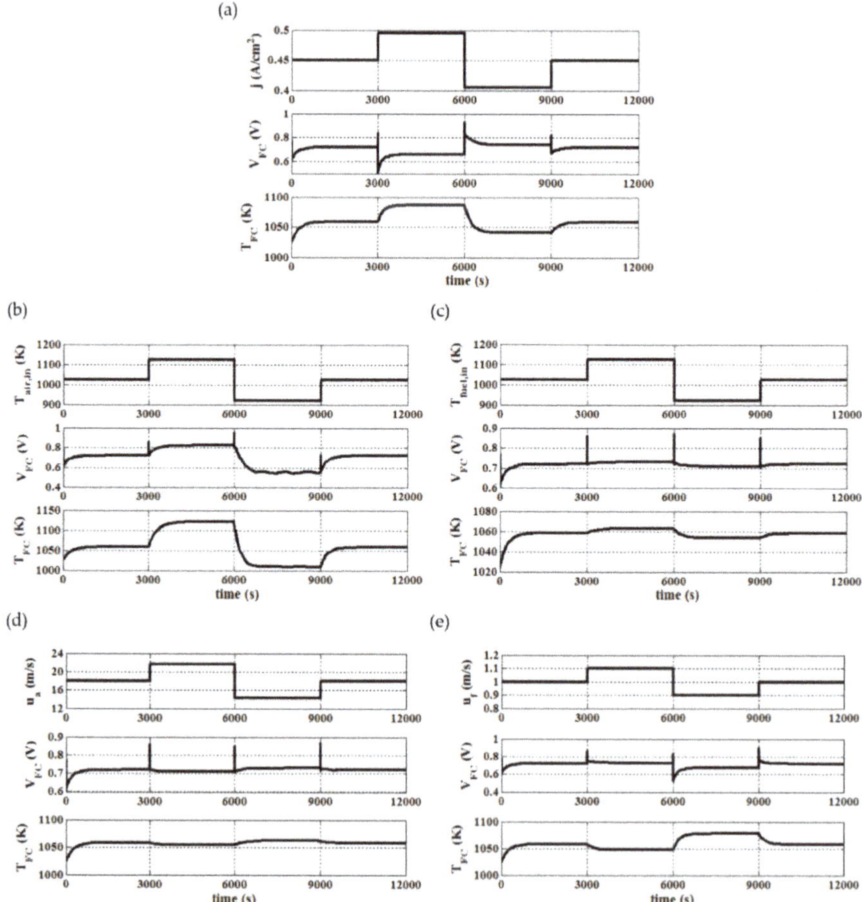


**Figure 2.** Voltage and cell temperature responses due to step changes in: (**a**) the current density ($j$); (**b**) the inlet temperature of air ($T_{\text{air,in}}$); (**c**) the inlet temperature of fuel ($T_{\text{fuel,in}}$); (**d**) the molar flow rate of air in terms of velocity ($u_a$); and, (**e**) the molar flow rate of fuel in terms of velocity ($u_f$).

### 5.2. Control of SOFC

In this part, the implementation of the ellipsoidal off-line robust MPC algorithm for LTV systems is presented and performed while using SeDuMi [46] and YALMIP [47]. The cell voltage ($V_{FC}$) is considered as the uncertainty parameter and it is assumed to be arbitrarily, varying in time within the indicated range. The lumped-parameter model of the SOFC that is represented by the nonlinear mass and energy balance ODEs is linearized, as follows:

$$\begin{aligned} \dot{\mathbf{x}} &= \mathbf{A}\mathbf{x} + \mathbf{B}\mathbf{u} \\ \mathbf{y} &= c\mathbf{x} \end{aligned} \tag{31}$$

where $\mathbf{x}$ is the state variables of the SOFC, i.e., the moles of each species in the fuel and air channels and the cell temperature, $\mathbf{u}$ is the inputs, such as the inlet molar flow rates of air and fuel, $\mathbf{A}$ and $\mathbf{B}$ are the matrices obtained from linearization, and $\mathbf{y}$ is the output variables.

Next, the linearized model is discretized while using an Euler first-order approximation in the discrete-time model expressed in Equation (32), with a sampling period of 5 s. Let $\bar{\mathbf{x}}(k) = \mathbf{x} - \mathbf{x}_{SS}$ and $\bar{\mathbf{u}}(k) = \mathbf{u} - \mathbf{u}_{SS}$, where the subscript *ss* denotes the corresponding variable at the steady-state condition, which results in the following:

$$\begin{aligned} \bar{\mathbf{x}}(k+1) &= \mathbf{A}(k)\bar{\mathbf{x}}(k) + \mathbf{B}(k)\bar{\mathbf{u}}(k) \\ \bar{\mathbf{y}}(k) &= \mathbf{c}\bar{\mathbf{x}}(k) \end{aligned} \tag{32}$$

An increase in cell temperature causes the material stresses, which is a potential problem, resulting in the anode and electrolyte material cracking. Additionally, the voltage must be controlled to make a high-efficiency SOFC. The objective is to control the cell temperature and the moles of methane at their desired values by manipulating the inlet molar flow rate of air and fuel with weighting matrices $\mathbf{Q}_1 = I$ and $\mathbf{R} = 0.1\, I$. In this study, the polytopic uncertainty model includes two vertices due to the existence of one uncertain parameter, $V_{FC}$. This parameter is randomly varied between 0.6 and 0.8 V in time.

Figure 3 shows the schematic diagram of the MIMO control system for the SOFC. The dashed line shows the off-line robust MPC algorithm in which the optimization problem is solved to obtain the corresponding state feedback gain, F. The solid line represents the measured on-line states at each sampling time and the corresponding state feedback control law, $\mathbf{u}$. The SOFC non-linear model is linearized (the state, input, and output variables are expressed in the deviation variables) and then discretized while using an Euler first-order approximation, which results in a discrete-time model. The uncertain parameters are implemented with the discrete-time model to generate the vertices sets. After the LTV system is obtained, it is implemented with the robust MPC algorithm. To obtain the gain $F$, the values of $Q_i$ and $Y_i$ are determined after optimization. Lastly, the state feedback control, $\mathbf{u}$, can be calculated and implemented in the control of the SOFC.

Figure 4 shows the closed-loop responses of the SOFC for the robust MPC based on the LTV systems. Figure 4a shows the closed-loop response of the cell temperature of the SOFC, whereas Figure 4b shows the closed-loop response of the moles of methane in the fuel channel. Figure 5a,b, respectively, show the control inputs of the SOFC, i.e., the inlet molar flow rates of air and fuel in velocity terms. Figure 6 shows the response of the cell voltage; the proposed MPC controller can indirectly drive the cell voltage to its desired value. The results show that when fuel cell temperature shifts from its steady-state value (1058 K), the mole fraction of methane has a slight change from the control action. The controller raises a small flow rate of fuel to reduce the temperature by enhancing the endothermic reforming reaction, which results in a slight increase in methane. In addition, the air flow is initially reduced to decrease the heat input. The robust MPC reduces the fuel flow while increasing the air flow to minimize its impact on controlled variables due to the increased methane. As a change in inputs could affect both of the controlled variables due to the inputs-outputs interaction, the MPC controller had to adjust both of the control inputs, which results in fluctuations in the initial control input profiles. The simulation results show that the proposed control algorithm obtains good results. The MPC controller for the LTV system can maintain the cell temperature and the moles of methane at their setpoints by manipulating the inlet molar flow rates of air and fuel, respectively.

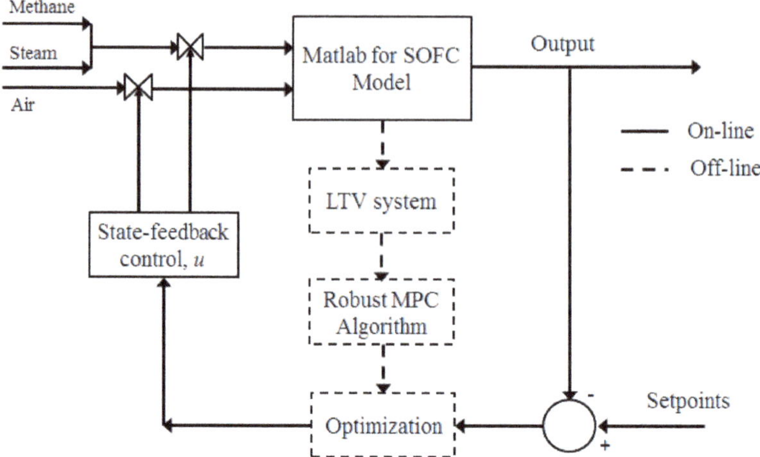


**Figure 3.** Schematic diagram of the control system for SOFC.

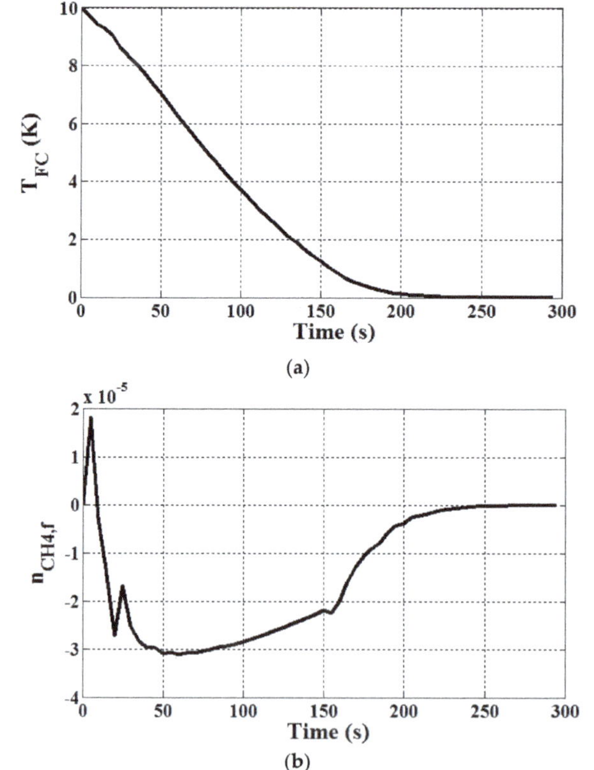


**Figure 4.** Closed-loop responses of SOFC: (**a**) the cell temperature of SOFC; and, (**b**) the moles of methane in fuel channel ($T_{FC}$ and $n_{CH4,f}$ are in a deviation form).

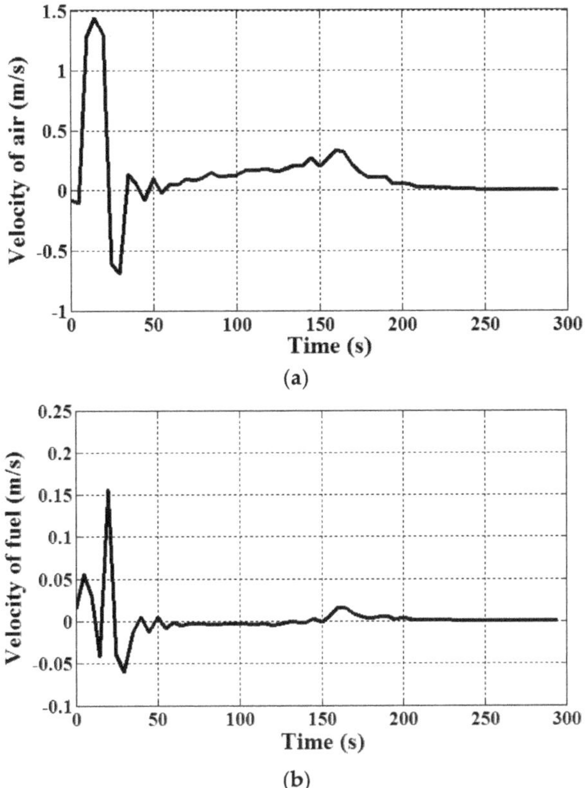


**Figure 5.** Control inputs of SOFC: (**a**) the inlet molar flow rate of air in terms of velocity; and, (**b**) the inlet molar flow rate of fuel in terms of velocity (velocities of air and fuel are in a deviation form).

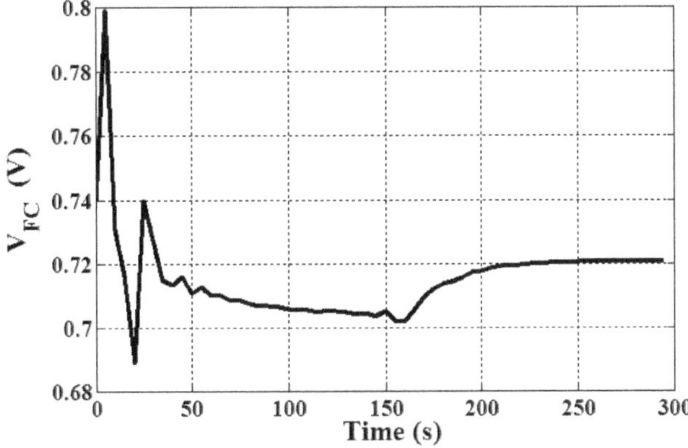


**Figure 6.** Closed-loop response of the cell voltage.

## 6. Conclusions

In this paper, an off-line robust MPC algorithm for a discrete-time LTV system with polytopic uncertainty while using the ellipsoidal invariant set was synthesized and designed for controlling a solid oxide fuel cell. The state feedback control law minimizing an upper bound on the worst-case objective function was implemented. The lumped-parameter model was employed to explain the SOFC's dynamic behavior and design the MPC controller. In the open-loop dynamic simulations, the inlet fuel and air temperature, and the current density are related to the fuel cell temperature and voltage. Regarding the performance of the SOFC control, the off-line robust MPC algorithm can guarantee the stability of the SOFC under the model uncertainty. The controller can keep the operating temperature and the moles of methane at their setpoints by manipulating the inlet molar flow rate of air and fuel. Consequently, the cell voltage also moves to its desired value.

**Author Contributions:** Conceptualization, N.C. and A.A.; methodology, N.C. and A.A.; validation, N.C.; formal analysis, N.C. and A.A.; investigation, N.C.; S.K.; Y.-S.C. and A.A.; writing—original draft preparation, N.C. and A.A.; writing—review and editing, S.K.; Y.-S.C. and A.A.; supervision, S.K.; Y.-S.C. and A.A.; project administration, A.A.; funding acquisition, A.A.

**Funding:** This research was funded by Chulalongkorn Academic Advancement into Its 2nd Century Project, Chulalongkorn University.

**Conflicts of Interest:** The authors declare no conflict of interest.

## References

1. Saedea, D.; Authayanun, S.; Patcharavorachot, Y.; Arpornwichanop, A. Effect of anode-cathode exhaust gas recirculation on energy recuperation in a solid oxide fuel cell-gas turbine hybrid power system. *Energy* **2016**, *94*, 218–232.
2. Atawi, I.E.; Kassem, A.M.; Zaid, S.A. Modeling, Management, and Control of an Autonomous Wind/Fuel Cell Micro-Grid System. *Processes* **2019**, *7*, 85. [CrossRef]
3. Zhao, Y.; Xia, C.; Jia, L.; Wang, Z.; Li, Y. Recent progress on solid oxide fuel cell: Lowering temperature and utilizing non-hydrogen fuels. *Int. J. Hydrog. Energy* **2013**, *38*, 16498–16517. [CrossRef]
4. Palomba, V.; Ferraro, M.; Frazzica, A.; Vasta, S.; Sergi, F.; Antonucci, V. Experimental and numerical analysis of a SOFC-CHP system with adsorption and hybrid chillers for telecommunication applications. *Appl. Energy* **2018**, *216*, 620–633. [CrossRef]
5. Fuqiang, W.; Lin, J.; Ziming, C.; Huaxu, L.; Jianyu, T. Combination of thermodynamic analysis and regression analysis for steam and dry methane reforming. *Int. J. Hydrog. Energy* **2019**, *44*, 15795–15810. [CrossRef]
6. Liao, C.H.; Horng, R.F. Experimental study of syngas production from methane dry reforming with heat recovery strategy. *Int. J. Hydrog. Energy* **2017**, *42*, 25213–25224. [CrossRef]
7. Lo Faro, M.; Trocino, S.; Zignani, S.C.; Italiano, C.; Vita, A.; Arico, A.S. Study of a solid oxide fuel cell fed with n-dodecane reformate. Part II: Effect of the reformate composition. *Int. J. Hydrog. Energy* **2017**, *42*, 1751–1757. [CrossRef]
8. Mustapha, N.A.; Sharuddin, S.S.; Zainudin, M.H.M.; Ramli, N.; Shirai, Y.; Maeda, T. Inhibition of methane production by the palm oil industrial waste phospholine gum in a mimic enteric fermentation. *J. Clean. Prod.* **2017**, *165*, 621–629. [CrossRef]
9. De Lorenzo, G.; Corigliano, O.; Lo Faro, M.; Frontera, P.; Antonucci, P.; Zignani, S.C.; Trocino, S.; Mirandola, F.A.; Aricò, A.S.; Fragiacomo, P. Thermoelectric characterization of an intermediate temperature solid oxide fuel cell system directly fed by dry biogas. *Energy Convers. Manag.* **2016**, *127*, 90–102. [CrossRef]
10. De Lorenzo, G.; Fragiacomo, P. Electrical and thermal analysis of an intermediate temperature IIR- SOFC system fed by biogas. *Energy Sci. Eng.* **2018**, *6*, 60–72. [CrossRef]
11. Triphob, N.; Wongsakulphasatch, S.; Kiatkittipong, W.; Charinpanitkul, T.; Praserthdam, P.; Assabumrungrat, S. Integrated methane decomposition and solid oxide fuel cell for efficient electrical power generation and carbon capture. *Chem. Eng. Res. Des.* **2012**, *90*, 2223–2234. [CrossRef]
12. Abdelkareem, M.A.; Tanveer, W.H.; Sayed, E.T.; Assad, M.; Allagui, A.; Cha, S.W. On the technical challenges affecting the performance of direct internal reforming biogas solid oxide fuel cells. *Renew. Sustain. Energy Rev.* **2019**, *101*, 361–375. [CrossRef]
13. Anderson, T.; Vijay, P.; Tade, M.O. An adaptable steady state Aspen Hysys model for the methane fuelled solid oxide fuel cell. *Chem. Eng. Res. Des.* **2014**, *92*, 295–307. [CrossRef]

14. Kupecki, J.; Papurello, D.; Lanzini, A.; Naumovich, Y.; Motylinski, K.; Blesznowski, M.; Santarelli, M. Numerical model of planar anode supported solid oxide fuel cell fed with fuel containing $H_2S$ operated in direct internal reforming mode (DIR-SOFC). *Appl. Energy* **2018**, *230*, 1573–1584. [CrossRef]
15. Pret, M.G.; Ferrero, D.; Lanzini, A.; Santarelli, M. Thermal design, modeling and validation of a steam-reforming reactor for fuel cell applications. *Chem. Eng. Res. Des.* **2015**, *104*, 503–512. [CrossRef]
16. Vijay, P.; Hosseini, S.; Tade, M.O. A novel concept for improved thermal management of the planar SOFC. *Chem. Eng. Res. Des.* **2013**, *91*, 560–572. [CrossRef]
17. Vijay, P.; Tade, M.O. An adaptive non-linear observer for the estimation of temperature distribution in the planar solid oxide fuel cell. *J. Process Control* **2013**, *23*, 429–443. [CrossRef]
18. Lu, X.; Li, T.; Bertei, A.; Cho, J.I.S.; Heenan, T.M.M.; Rabuni, M.F.; Li, K.; Brett, D.J.L.; Shearing, P.R. The application of hierarchical structures in energy devices: New insights into the design of solid oxide fuel cells with enhanced mass transport. *Energy Environ. Sci.* **2018**, *11*, 2390–2403. [CrossRef]
19. Chaisantikulwat, A.; Diaz-Goano, C.; Meadows, E.S. Dynamic modeling and control of planar anode-supported solid oxide fuel cell. *Comput. Chem. Eng.* **2008**, *32*, 2365–2381. [CrossRef]
20. Madani, O.; Das, T. Feedforward based transient control in solid oxide fuel cells. *Control Eng. Pract.* **2016**, *56*, 86–91. [CrossRef]
21. Lotfi, N.; Zomorodi, H.; Landers, R.G. Active disturbance rejection control for voltage stabilization in open-cathode fuel cells through temperature regulation. *Control Eng. Pract.* **2016**, *56*, 92–100. [CrossRef]
22. Li, Y.H.; Choi, S.S. An analysis of the control and operation of a solid oxide fuel-cell power plant in an isolated system. *IEEE Trans. Energy Conver.* **2005**, *20*, 381–387. [CrossRef]
23. Aguiar, P.; Adjiman, C.S.; Brandon, N.P. Anode-supported intermediate-temperature direct internal reforming solid oxide fuel cell. II: Model-based dynamic performance and control. *J. Power Sources* **2005**, *147*, 136–147. [CrossRef]
24. Stiller, C.; Thorud, B.; Bolland, O.; Kandepu, R.; Imsland, L. Control strategy for a solid oxide fuel cell and gas turbine hybrid system. *J. Power Sources* **2006**, *158*, 303–315. [CrossRef]
25. Zhang, X.W.; Chan, S.H.; Ho, H.K.; Li, J.; Feng, Z. Nonlinear model predictive control based on the moving horizon state estimation for the solid oxide fuel cell. *Int. J. Hydrog. Energy* **2008**, *33*, 2355–2366. [CrossRef]
26. Gutierrez, G.; Ricardez-Sandoval, L.A.; Budman, H.; Prada, C. An MPC-based control structure selection approach for simultaneous process and control design. *Comput. Chem. Eng.* **2014**, *70*, 11–21. [CrossRef]
27. Wu, F. LMI-based robust model predictive control and its application to an industrial CSTR problem. *J. Process Control* **2001**, *11*, 649–659. [CrossRef]
28. Tahir, F.; Jaimoukha, I.M. Robust feedback model predictive control of constrained uncertain systems. *J. Process Control* **2013**, *23*, 189–200. [CrossRef]
29. Kothare, M.V.; Balakrishnan, V.; Morari, M. Robust constrained model predictive control using linear matrix inequalities. *Automatica* **1996**, *32*, 1361–1379. [CrossRef]
30. Bahakim, S.S.; Ricardez-Sandoval, L.A. Simultaneous design and MPC-based control for dynamic systems under uncertainty: A stochastic approach. *Comput. Chem. Eng.* **2014**, *63*, 66–81. [CrossRef]
31. Manthanwar, A.M.; Sakizlis, V.; Pistikopoulos, E.N. Robust parametric predictive control design for polytopically uncertain systems. In Proceedings of the American Control Conference, Portland, OR, USA, 8–10 June 2005; pp. 3994–3999.
32. Bumroongsri, P.; Kheawhom, S. An off-line robust MPC algorithm for uncertain polytopic discrete-time systems using polyhedral invariant sets. *J. Process Control* **2012**, *22*, 975–983. [CrossRef]
33. Pannocchia, G. Robust model predictive control with guaranteed setpoint tracking. *J. Process Control* **2004**, *14*, 927–937. [CrossRef]
34. Kouramas, K.; Varbanov, P.S.; Georgiadis, M.C.; Klemes, J.J.; Pistikopoulos, E.N. Explicit/multi-parametric model predictive control of a solid oxide fuel cell. In Proceedings of the 21st European Symposium on Computer Aided Chemical Engineering (ESCAPE21), Halkidiki, Greece, 29 May–1 June 2011; pp. 773–777.
35. Wan, Z.; Kothare, M.V. Robust output feedback model predictive control using off-line linear matrix inequalities. *J. Process Control* **2002**, *12*, 763–774. [CrossRef]
36. Pirkandi, J.; Ghassemi, M.; Hamedi, M.H.; Mohammadi, R. Electrochemical and thermodynamic modeling of a CHP system using tubular solid oxide fuel cell (SOFC-CHP). *J. Clean. Prod.* **2012**, *29–30*, 151–162. [CrossRef]
37. Yang, Y.; Du, X.; Yang, L.; Huang, Y.; Xian, H. Investigation of methane steam reforming in planar porous support of solid oxide fuel cell. *Appl. Therm. Eng.* **2009**, *29*, 1106–1113. [CrossRef]

38. Achenbach, E.; Riensche, E. Methane/steam reforming kinetics for solid oxide fuel cells. *J. Power Sources* **1994**, *52*, 283–288. [CrossRef]
39. Xi, H.; Varigonda, S.; Jing, B. Dynamic modeling of a solid oxide fuel cell system for control design. In Proceedings of the American Control Conference, Baltimore, MD, USA, 30 June–2 July 2010; pp. 423–428.
40. Murshed, A.M.; Huang, B.; Nandakumar, K. Control relevant modeling of planar solid oxide fuel cell system. *J. Power Sources* **2007**, *163*, 830–835. [CrossRef]
41. Ni, M.; Leung, M.K.H.; Leung, D.Y.C. Micro-scale modeling of oxide fuel cells with micro-structurally graded electrodes. *J. Power Sources* **2007**, *168*, 369–378. [CrossRef]
42. Patcharavorachot, Y.; Arpornwichanop, A.; Chuachuensuk, A. Electrochemical study of a planar solid oxide fuel cell: Role of support structures. *J. Power Sources* **2008**, *177*, 254–261. [CrossRef]
43. Liu, R.W. Convergent systems. *IEEE Trans. Autom. Control* **1968**, *13*, 384–391.
44. Boyd, S.; Ghaoui, L.E.; Feron, E.; Balakrishnan, V. *Linear Matrix Inequalities in System and Control Theory*; SIAM Studies in Applied Mathematics: Philadelphia, PA, USA, 1994.
45. Georgis, D.; Jogwar, S.S.; Almansoori, A.S.; Daoutidis, P. Design and control of energy integrated SOFC systems for in situ hydrogen production and power generation. *Comput. Chem. Eng.* **2011**, *35*, 1691–1704. [CrossRef]
46. Sturm, J.F. Using Sedumi 1.02, a MATLAB toolbox for optimization over symmetric cones. *Optim. Methods Softw.* **1999**, *11*, 625–653. [CrossRef]
47. Lofberg, J. YALMIP: A toolbox for modeling and optimization in MATLAB. In Proceedings of the IEEE International Symposium on Robotics and Automation, New Orleans, LA, USA, 2–4 September 2004; pp. 284–289.

*Article*

# Simulation of Solid Oxide Fuel Cell Anode in Aspen HYSYS—A Study on the Effect of Reforming Activity on Distributed Performance Profiles, Carbon Formation, and Anode Oxidation Risk

**Khaliq Ahmed [1], Amirpiran Amiri [2,*] and Moses O. Tadé [1]**

[1] Department of Chemical Engineering, Curtin University, Bentley 6102, Australia; Khaliq.Ahmed@curtin.edu.au (K.A.); m.o.tade@curtin.edu.au (M.O.T.)

[2] Energy and Bioproducts Research Institute (EBRI), School of Engineering and Applied Science, Aston University, Birmingham B4 7ET, UK

* Correspondence: a.p.amiri@aston.ac.uk

Received: 11 November 2019; Accepted: 24 February 2020; Published: 27 February 2020

**Abstract:** A distributed variable model for solid oxide fuel cell (SOFC), with internal fuel reforming on the anode, has been developed in Aspen HYSYS. The proposed model accounts for the complex and interactive mechanisms involved in the SOFC operation through a mathematically viable and numerically fast modeling framework. The internal fuel reforming reaction calculations have been carried out in a plug flow reactor (PFR) module integrated with a spreadsheet module to interactively calculate the electrochemical process details. By interlinking the two modules within Aspen HYSYS flowsheeting environment, the highly nonlinear SOFC distributed profiles have been readily captured using empirical correlations and without the necessity of using an external coding platform, such as MATLAB or FORTRAN. Distributed variables including temperature, current density, and concentration profiles along the cell length, have been discussed for various reforming activity rates. Moreover, parametric estimation of anode oxidation risk and carbon formation potential against fuel reformation intensity have been demonstrated that contributes to the SOFC lifetime evaluation. Incrementally progressive catalyst activity has been proposed as a technically viable approach for attaining smooth profiles within the SOFC anode. The proposed modeling platform paves the way for SOFC system flowsheeting and optimization, particularly where the study of systems with stack distributed variables is of interest.

**Keywords:** SOFC; simulation; internal reforming; anode oxidation; carbon formation

## 1. Introduction

Fuel cells convert the chemical energy in fuel directly into electricity and heat, without combustion, leading to high efficiencies with low or even zero emissions. The SOFC is becoming a mature technology and can make the commercial breakthrough if cost targets can be met by achieving cost reductions through volume manufacturing, improved lifespan/performance, and lower cost materials [1–3]. Research and development in the last twenty years have led to significant advances in all areas of the technology including cell, seal, interconnect, and stack design, as well as peripheral components and the entire balance of plant (BoP) [4,5]. Manufacturing achievements have led to defect identification and minimization, quality control, and scale-up of stack components and the entire stack assembly manufacture.

The SOFC system is appropriate to operate on a pipeline fuel such as reticulated natural gas with its well-established supply infrastructure throughout the world. For such a fuel, minimal fuel processing is required, which includes desulphurization of the fuel to remove sulphur compounds that are naturally present in the hydrocarbon fuel source and those that are added as odorants to meet legislative requirements such as for natural gas, propane, and LPG for domestic applications. The preprocessing also includes a level of conversion of the hydrocarbon fuel, conventionally known as prereforming, which functions to convert the hydrocarbon feed such as natural gas to a hydrogen-rich mixture or a methane-rich mixture depending on the type of anode in the SOFC stack, i.e., noninternal reforming or internal reforming type. For an internal reforming type SOFC, where methane can be converted by steam reforming on the anode, it suffices to prereform the fuel to a level where all higher hydrocarbons ($C_{1+}$) are converted leading to a mixture of methane, hydrogen, and carbon oxides with little or no conversion of methane [6,7]. For a noninternal reforming type SOFC, all hydrocarbon components including methane need to be fully converted to a mixture of hydrogen and carbon oxides. Owing to its high electrical efficiency, the SOFC technology results in reduced emissions of $CO_2$ and is practically noise free. Furthermore, it is free of $NO_x$ emissions due to its relatively low operating temperatures. The SOFC system is particularly attractive as a combined heat and power generation (CHP) system, since the waste heat generated can be used to supply heat to a hot-water system which can be interfaced to the SOFC system [8].

A system-level flowsheet model of the SOFC system including the complete BoP is a useful platform for simulating the performance of the plant and for sizing of individual components of the BoP. Commercially available process simulation software such as Aspen Plus or Aspen HYSYS, PRO/II, etc., contains extensive thermodynamic and physical properties database and includes in-built modules for a number of components which are commonly used in a process plant, such as heat-exchangers of various types, reactors of various types, compressors, pumps, valves, separating columns, tanks, mixers, etc. It allows for energy optimization via heat and work integration of system components. However, it does not include a module for fuel cell reactions, i.e., it cannot directly account for the electrochemical reactions involving ions and electrons. There are two approaches for modeling SOFC-based systems with commercial process simulators. In one approach, the SOFC model is developed in a separate platform such as FORTRAN, VB, C++, MATLAB, etc. and then linked to the process simulator [9–15]. In another approach, the SOFC reactions are modeled using the equilibrium reactor module GIBBS [16,17]. Anderson et al. [17] modeled the SOFC as a combination of an isothermal plug flow reactor (PFR) module, to account for methane reforming kinetics on the anode, and a GIBBS reactor for the fuel cell reactions of hydrogen and CO oxidation. However, this was not a system-level model and focused on reactions and mass transport processes at the cell level. Using established theoretical and/or empirical correlations from literature, they tested the validity of their model by comparing their simulation results with those of others reported in the literature. Two main drawbacks of this work are the assumptions of isothermal conditions for internal reforming and use of a GIBBS reactor for the fuel cell reactions. In a real system the SOFC stack does not operate in an isothermal mode. There are two opposing contributions to stack temperature profiles in the case of an internal reforming anode. The endothermic steam reforming reaction absorbs heat from the gas stream which results in cooling of the stack and the fuel cell reaction(s) release heat which results in heating of the stack; the net effect is determined by the extents of these reactions.

A conversion reactor is more appropriate for representing the fuel oxidation reactions by setting the percent conversion equal to fuel utilization. An equilibrium approach using the GIBBS reactor does not allow setting of the reaction conversion to match fuel utilization. In this work, the internal reforming of methane via steam reforming and the accompanying water-gas shift (WGS) reaction is modeled via the PFR module with the kinetic expressions from literature [18,19] and the fuel cell reactions are modeled using the conversion reactor where the conversion is linked to the fuel utilization value calculated in a spreadsheet block. Another feature of the current work is that unlike the work of Anderson et al. [17], where the PFR is modeled as an isothermal reactor, in this work the energy stream of the PFR is linked to the cell in the spreadsheet block which calculates the heat generated by the fuel cell reaction and is available as reaction heat in a direct internal reforming SOFC. The axial temperature profile created in the PFR is therefore, representative of the temperature profile on an SOFC anode with direct internal reforming, as the coupling of the endothermic methane steam reforming (MSR) reaction and accompanying mildly exothermic WGS reaction with heat available from the fuel cell reaction is appropriately captured with this approach. The corresponding composition profile under current load cannot be generated within the PFR module as this module only works with kinetic schemes and it is not possible to add the fuel cell reactions to the PFR as conversion reactions based on fuel utilization. An option to include the reaction kinetics of the electrochemical oxidation of hydrogen is also not available in the software. Nevertheless, the reaction extents and accompanying heat exchanges can be calculated in the spreadsheet module and linked to the PFR module. Firstly, this allows generation of open-circuit composition profiles of the internally reformed gas which sets the boundary for the Nernst voltage profile under load, after accounting for the extents of the fuel cell reactions. Secondly, the current density and composition profiles can be calculated within the spreadsheet block using appropriate correlations.

Previous work [9–13,16] largely focused on the issues that can predict and improve the fuel cell operation in terms of current generation and voltage losses. For instance, the effect of air flow rate, steam to carbon ratio (S/C), current density, fuel utilization (Uf), inlet temperatures, or operating pressure have been extensively investigated. By contrast, in this work we have mainly focused on an analysis of processes that significantly affect anode performance and lifetime and consequently impact on the SOFC system as a whole. We analyzed anode performance for various levels of reforming activity. Three cases are considered: (i) Full reforming activity, (ii) 1/3rd reforming activity, and (iii) 1/6th reforming activity. Reduced reforming activity may be the result of engineered design [20] or may result from progressive degradation of the anode from poisoning or sintering due to nickel coarsening over the useful life of the stack [4], which extends the reaction zone and requires more of the anode segment from the leading edge to fully convert methane. Reforming kinetics reported by Ahmed and Föger [18] and WGS reaction kinetics reported by Tingey [19] were employed as the reaction rate details for PFR, leading to 1D pseudo-homogeneous results. The three different levels of activity were assigned by reducing the Arrhenius factor in the rate expression by the reduction factors 0.33 (~1/3rd) and 0.67 (~1/6th). In physical terms this signifies loss of reforming activity by poisoning or sintering. For these levels of activity, we assess the anode oxidation risk and carbon formation potential on the anode, both of which have severe life-limiting consequences on the anode [7].

This paper contributes to the SOFC research considering its two novel contents including; (i) the novel simulation methodology: The simulation approach proposed for complicated internal reforming SOFC process offers simplicity and calculation speed without compromising the internal operations details. This is of particular interest for SOFC system modeling and design where several operational concepts including heat/mass transfer and electrochemical and fuel reformation reactions interactively occur at wide time and length scales. (ii) The understanding of distributed reformation potentials in controlling SOFC performance profiles. The incremental reformation process is demonstrated to be a promising strategy to moderate the undesirable gradients of SOFC internal profiles. This is promising to achieve higher homogeneity in temperature and concertation profiles inside the SOFC stack that subsequently offers enhanced current and voltage profiles. This is crucially important for SOFC efficiency and durability. In this paper, we demonstrate internal fuel reformation as an opportunity not only for heat integration and external reformer cost reduction but also for thermal management goals that eventually results in fuel cell longevity.

## 2. Simulation Methodology

The SOFC stack is simulated in the following way: Internal reforming of methane on the SOFC anode is modeled in a PFR module, using the reforming kinetics of Ahmed and Föger [18] and the WGS reaction kinetics of Tingey [19]. The electrochemical conversion of hydrogen and carbon monoxide are modeled as chemical conversions in a conversion reactor module. The associated electrical aspects including cell voltage, air utilization, and fuel utilization at a given operating current are calculated in a spreadsheet tool of Aspen HYSYS. The spreadsheet block in Aspen HYSYS is essentially an Excel-based spreadsheet with features of exporting data to and importing data from other modules of the flowsheet. This provides facility for post-processing of the calculated or entered values of the process variables. Computations of cell voltage entailing calculations of electrical losses, both ohmic and overpotential using empirical correlations [21] and calculation of Nernst voltage from the concentrations of the reacted gas are carried out in the spreadsheet, with compositions imported from the PFR module. Similarly, the sum of electrical losses and entropy change, after allowing for losses to the surroundings from the stack, is calculated and this value is exported to the energy stream linked to the PFR, as heat available for direct internal reforming.

The average stack operating temperature is obtained by a trial-and-error method. An average stack temperature is assumed for calculating Nernst voltage, operating voltage, and electrical losses. The assumed temperature is then compared to the value returned by the PFR, which simulates internal reforming and utilizes heat generated from the fuel cell reaction to compute the temperature profile within the fuel cell and the composition profile of the internally reformed fuel. The amount of heat generated, i.e., the electrical and entropy losses are calculated at the assumed average temperature. These steps are iterated until agreement is reached between the assumed average cell temperature and the calculated average cell temperature.

The composition of utilized gas is obtained by matching the degree of conversion in the conversion reactor, with the fuel utilization level calculated in the spreadsheet, based on fuel flow rate and operating current. Since the electrochemical conversion of $H_2$ and CO are modeled as combustion reactions, there is a temperature rise in the reactor due to the exothermicity of the combustion reactions. Since the exothermicity of the fuel cell reactions have already been accounted for by calculating the electrical losses and entropy change of the fuel cell reaction and entering this value as heat input to the internal reforming PFR, temperature rise in the conversion reactor is suppressed by use of the HYSYS object SET, to set the temperature of the stream leaving the conversion reactor to be the same as the temperature of the stream leaving the PFR.

Figure 1 and Table 1 show the integration of the modules representing the anode operation and the set of equations used in the calculation of all distributed variables within the stack—temperature, current density, and composition of all chemical species on the anode side.

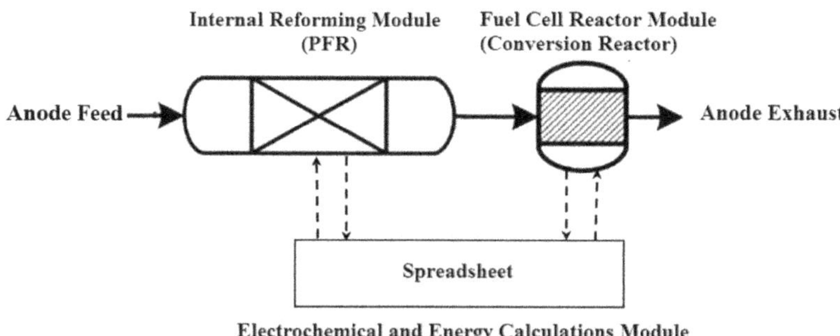


**Figure 1.** Schematic of integration of the internal reforming reaction modules and the fuel cell reactor module.

The reactions taking place in the reaction modules, shown in Figure 1, are as follows. Two parallel reactions take place in the internal reforming module (PFR) including MSR and WGS, presented by Equations (1) and (2), respectively:

$$CH_4 + H_2O = CO + 3H_2 \tag{1}$$

$$CO + H_2O = CO_2 + H_2 \tag{2}$$

Two parallel electrochemical reactions take place in the fuel cell reactor module (conversion reactor). The main reaction is hydrogen oxidation (Equation (3)) that may occur with the CO oxidation (Equation (4)), simultaneously.

$$H_2 + 0.5O_2 = H_2O \tag{3}$$

$$CO + 0.5O_2 = CO_2 \tag{4}$$

**Table 1.** Modeling approach for solid oxide fuel cell (SOFC) performance approximation.

| Equations/Parameters | Comment |
|---|---|
| $\{Uf, \dot{m}_{Fuel}\} = Given$ | Constant fuel utilization. |
| $I^{av} = \dot{m}_{Fuel} Uf(nF)$ | Average current drawn based on given fuel rate and desired fuel utilization. |
| $Q_{elec} = (I^{av}/nF)(\Delta H^r) = \dot{m}_{Fuel} Uf(\Delta H^r)$ | Heat of electrochemical reaction for given fuel consumption, calculated in the spreadsheet block. |
| $\Delta T = \left(\frac{Q_{elec}}{(mC_p)_{mix}}\right)$ | Temperature change due to heat release from fuel oxidation. |
| $C_{x,OCO}(l), T_{OCO}(l)$ $x = \{H_2, H_2O, CO, CO_2, CH_4\}$, | Open-circuit operation (OCO) composition and temperature profiles from the internal reforming module (PFR) based on MSR and WGS reaction kinetics. |
| $T_{CCO}(l) = T_{OCO}(l) + \Delta T$ | Temperature profile for closed-circuit operation from corresponding open-circuit operation data. |
| $C_{x,CCO}(l) \cong f(C_{x,OCO}(l), Uf)$ | Closed-circuit operation (CCO) composition profiles using fuel utilization. |
| $I_{CCO}(l) \cong g(T_{CCO}(l), C_{H_2,OCO}(l))$ | Current density profile estimation from temperature distribution and open-circuit hydrogen concentration profile. |
| $r_{MSR} = k_o \exp(-E_a/RT)(p_{CH_4})^{\alpha}(p_{H_2O})^{\beta}$ | Kinetics of MSR; $k_o$ = 8542 mol/($m^2 bar^{1/2}$s), α = 0.85, β = −0.35, and $E_a$ = 95 kJ/mol [18]. |
| $r_{RWGS} = 7.6 \times 10^4 \exp(-39.2/RT) p_{H_2}{}^{0.333} p_{CO_2} + 1.2 \times 10^{13} \exp(-78/RT) p_{H_2}{}^{0.5} p_{CO_2}$ | Kinetics of the reverse WGS reaction reported by Tingey [19]. |
| $V = E^{OCV} - \eta - IR_{\Omega}$ | The cell voltage was calculated by deducting the voltage losses from the open-circuit Nernst voltage. |
| $E^{OCV} = E^{\circ} + \frac{RT}{nF} \ln\left(\frac{p_{H_2} \cdot p_{O_2}{}^{0.5}}{p_{H_2O}}\right)$ | Open-circuit Nernst voltage. |
| $\eta = (1/a) I^y U_f{}^z \exp(E_b/RT)$ | Overpotential losses were calculated using an empirical equation [21]; a = 4.43, y = 0.77, z = -0.15, and $E_b$ = 10,560 kJ/mol, determined by regression analysis of experimental measurement of cell overpotential at various current densities and fuel utilization at the three temperatures 750, 800, and 850 °C. |
| $R_{\Omega} = (t_e/A)\exp(B/T)$ | The cell ohmic resistance is independent of fuel utilization as expected [21], where $t_e$ is the electrolyte thickness (μm) and A = 21,428 μm/Ω and B = 7776 K are constants, with their values determined empirically [21]. |

## 3. Results and Discussion

The base case is a system operating at 65% fuel utilization. Input parameters for the base case simulation and also estimated results returned by the flowsheet calculations are presented in Tables 2 and 3, respectively.

**Table 2.** Model input parameters for the base case simulation.

| Variable | Value |
|---|---|
| NG Flow Rate | 5.4 SLM |
| Air Flow Rate | 300 SLM |
| Air Inlet Temperature to Stack | 810 °C |
| Fuel Utilization | 65% |
| S/C Ratio (at reformer inlet) | 2.25 |
| Fuel Inlet Temperature to Stack | 810 °C |
| *Note*: SLM—Standard litres per minute, at 0 °C, 1 atm | |

**Table 3.** Estimated SOFC performance results.

| Variable | Value |
|---|---|
| Net Power | 1.3 kW |
| Net Electrical Efficiency | 38.6% |
| Stack Operating Current | 40.7 A |
| Operating Voltage | 0.67 V |
| Air Utilization | 11.4% |
| Average Stack Temperature | 792.7 °C |
| Stack Exhaust Temperature | 847.4 °C |
| Nernst Voltage (Open-Circuit) | 0.993 V |
| Nernst Voltage (65% Uf) | 0.900 V |
| Overpotential Losses | 0.173 V |
| Ohmic Resistance | 1.547 ohm-cm$^2$ |
| Methane Slip | 0.0% |

### 3.1. Distributed Profiles

A high degree of internal reforming of methane is desirable as it reduces the prereformer size/cost and contributes to the cooling of the cell and therefore, reduces the cooling air flow requirement. However, the internal fuel reformation process may undesirably cause higher gradients in the distributed variables profiles. Stack operation homogeneity is of crucial importance from both efficiency and lifetime viewpoints. Figure 2 shows when the kinetics of reforming is too rapid, the cell temperature drops sharply near the cell inlet preventing an even distribution of current density and temperature. The steep temperature gradient is caused by the cooling impact associated with the endothermic steam reforming of methane on the anode. Figure 3 shows the reforming in this case is so fast that it is completed within the first 40% of the cell. Clearly, the maximum temperature gradient can be reduced if the sharp drop in temperature can be avoided, improving all temperature-dependent profiles including species concentration, current density, and overpotentials distributions, all of which have interacting effects. An effective way of achieving this objective is by using catalysts with lower overall reforming activity in anode and/or designing an anode with progressively increasing local activity along the cell length. A reduction of the anode activity to 1/3rd of that of the fast reforming anode results in a relatively uniform temperature profile as shown in Figure 2.

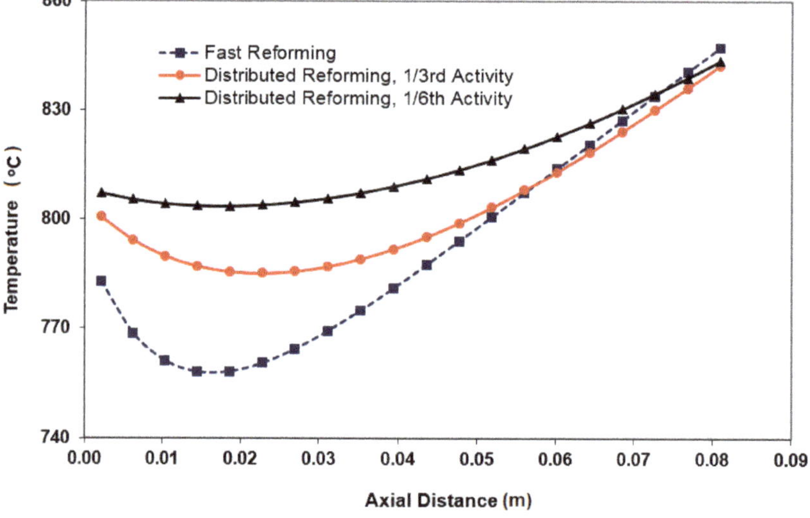


**Figure 2.** Temperature profiles for various levels of reforming activity.

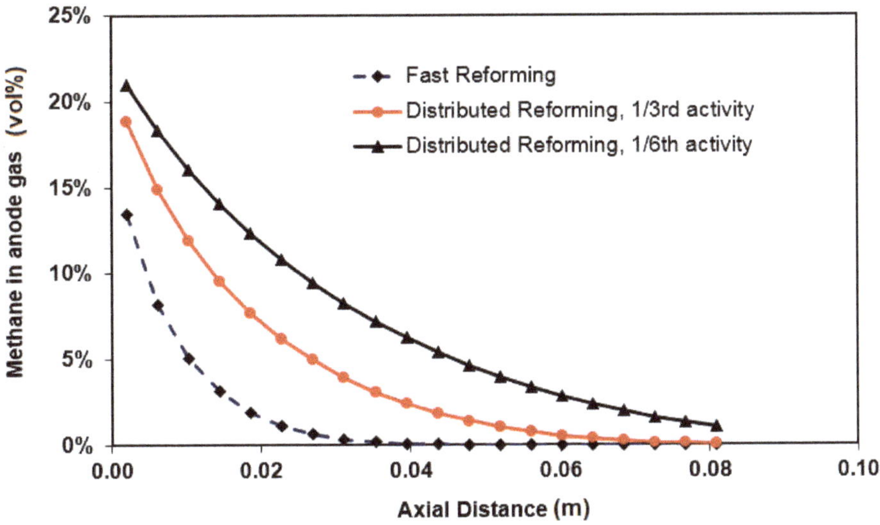


**Figure 3.** Methane concentration profiles for different reforming activity.

A key technical question is whether the fuel reformation should be fully accomplished in the first half of the cell length. If not, there are significant opportunities to design anode reforming activity targeting a more homogenous operation. Accordingly, the objective of achieving smoother temperature profiles in the operating cell must also be assessed from another viewpoint. The activities are compared to the basis case activity (fast reformation). For instance, 1/3rd activity means 66% less activity (lower rate) compared to the basis case. We have adjusted this via E (activation energy value) adjustment in the simulations. With activities lower than the basis case, the catalyst still may offer sufficient reformation as the methane consumption (Figure 3) and hydrogen generation profiles show. The speed of reformation, however, is declined while it is compensated by residence time (axial length). This is the distributed conversion in contrast to the sharp conversion seen in basis case. To further reduce the temperature differential, an anode with a reforming activity 1/6th of that of the base case fast reforming anode is considered. This results in a more uniform temperature profile, Figure 2, but comes at a cost of increasing methane slip as shown in Figure 3. While a more homogenous electrochemical reactor is achievable under distributed/progressive reforming conditions, an immediate concern relevant to the reduced activity is the possibility of reduction in the generated current due to less local $H_2$ availability in the first 30%–40% of the cell length. In order to assess this rigorously, the overall current produced over the cell surface must be calculated via integration of the local currents. Since no variation of current in cell width direction is assumed, that is a reasonable assumption for co-flow cell, two-dimensional integration can be replaced by one dimensional integration over the cell length. In such a case, therefore, the modeling approach proposed in this paper suffices. The surface under current density profiles in Figure 4 compares the total current production under different reforming approaches. Even though current production considerably drops in the cell inlet region when fast reforming occurs, the overall current production variation for fast and slow reformation activities can be reasonably ignored, indicating that cell efficiency will not be compromised for homogeneity.

Methane reforming profile (Figure 3) gives an indication of how the anode exhaust stream might be post-processed. In a fast reforming method, methane is completely consumed inside the stack leaving anode tail gas mainly including, hydrogen, CO, and a significant amount of steam. In such a case anode gas recycling is an appropriate process strategy. For slow/distributed fuel reforming, anode tail gas might contain some methane and less hydrogen and CO compared to the fast reforming. This can be understood by interpretation of methane and current profiles simultaneously. As current generations are equal, the $H_2$ consumption in all cases are almost the same. For a given methane rate at inlet, therefore, higher methane fraction in cell outlet indicates lower amount of hydrogen and CO. This becomes even more considerable for an activity as low as 1/6th of the base case. Therefore, an after-burner can be designed in the system to achieve high quality heat from the tail gas. It may be concluded that the progressive fuel reforming offers some advantage for a CHP system in which both high quality heat and homogenous stack performance are desirable.

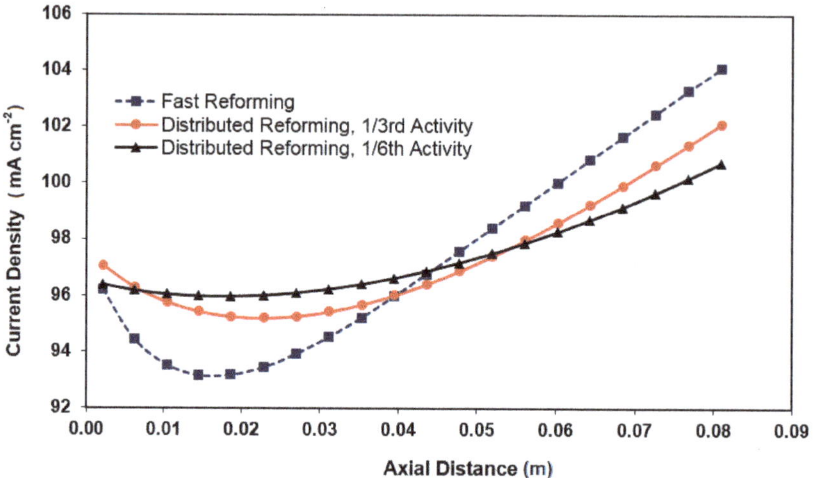


**Figure 4.** Current density profiles for different levels of reforming activity and Uf = 65%.

Stack exhaust temperature is reduced with progressive reforming activity compared to fast reforming, but the stack temperature profile for most of the anode length is higher as shown in Figure 2. While the latter is expected due to less steeper cooling along the cell, the reduction in stack exhaust temperature can be explained from a consideration of the associated current density profile. As shown in Figure 4, the current density profile is steepest for the fast reforming case resulting in higher joule heating effect. A lower joule heating effect results in lower increase in temperatures for the distributed reforming cases, to the point that the temperature at the end of the cell is lower for distributed reforming, even though the temperatures near the inlet are higher. A more uniform current density profile results in the case of lower reforming activity as expected which is beneficial from the point of stress reduction.

### 3.2. Anode Oxidation

The SOFC anode is susceptible to oxidation by steam in the reaction mixture according to Equation (5):

$$Ni + 2H_2O = Ni(OH)_2 + H_2 \tag{5}$$

In this paper, anode oxidation risk is determined on the basis of industrial experience [22] with nickel-based catalysts, where it reported finding that steam-to-hydrogen ratios greater than 6–8 increases the risk of nickel oxidation in nickel-based steam reforming catalysts. Analyses, therefore, are based on the local partial pressures as a characteristic indicator rather than estimation of rate for reaction 5. Distributed profiles show that the risk of anode oxidation is particularly high at high fuel utilization, where the partial pressure of steam in the reaction mixture is high, i.e., at high $p_{H2O}/p_{H2}$. Anode oxidation risk profile along the length of the cell is shown in Figure 5 for 65% and 75% fuel utilization with three levels of reforming activity. Anode oxidation risk is highest at the anode inlet for internal reforming anodes where enough hydrogen has not been generated (Figure 6). The risk increases with slow reforming activity. At higher utilization, more hydrogen is utilized, further lowering the $H_2$ content and increasing the $H_2O$ content, thereby increasing the risk of anode oxidation compared to lower utilization. In such a case, anode tail gas recycle may worsen the situation by introducing more steam upstream. This is an additional reason why tail gas in the slow internal reforming case is recommended to be burnt rather than being recycled. In general, anode gas recycling may change the $p_{H2O}$ and $p_{H2}$ balance in favour of steam.

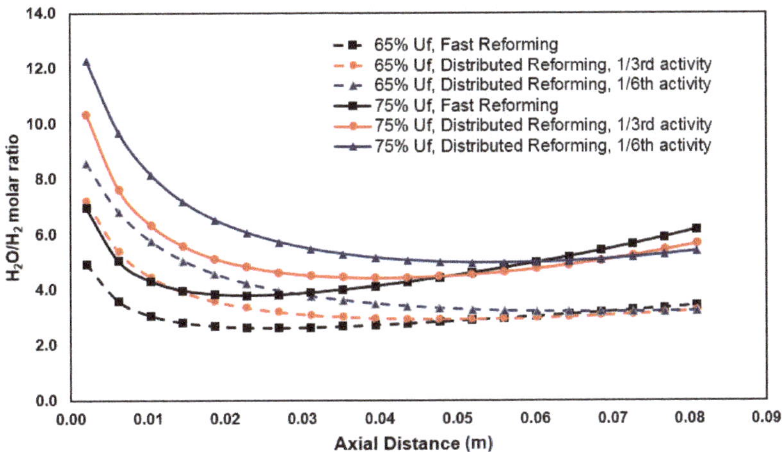


**Figure 5.** Anode oxidation risk profile for different reforming activity at 65% and 75% fuel utilization.

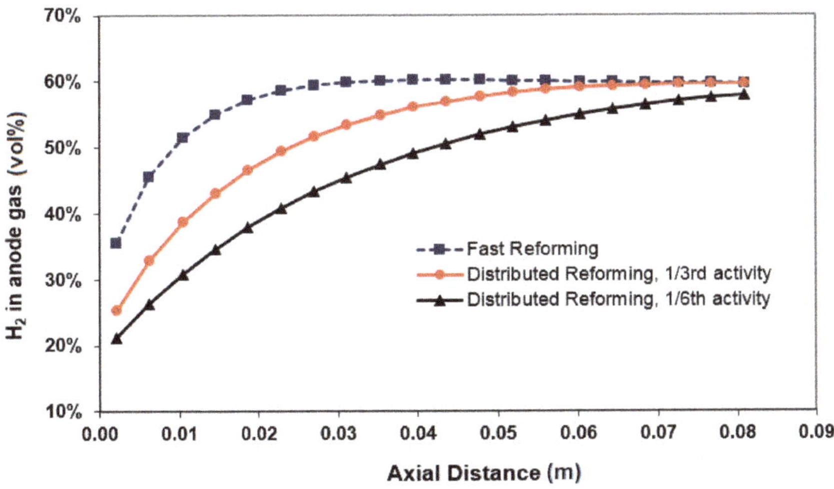


**Figure 6.** Open-circuit hydrogen concentration profiles for different reforming activity.

### 3.3. Carbon Formation

Carbon formation on the SOFC anode, as a challenging issue for SOFC degradation, might occur principally via two routes: Boudouard reaction and methane cracking. In this work, carbon formation activity is calculated based on the thermodynamical equilibrium estimated using the local temperature and compositions of the relevant species participating in the respective reactions. Carbon may form by cracking or dissociation of a methane molecule into carbon and hydrogen molecules according to the reaction in Equation (6):

$$CH_4 = C + 2H_2 \tag{6}$$

For this reaction, the carbon activity for dissociation of a methane ($a_c^d$) can be calculated by using Equation (7):

$$a_c^d = K_c^d \frac{p_{CH_4}}{\left(p_{H_2}\right)^2} \tag{7}$$

The Boudouard or carbon disproportionation reaction can be presented as Equation (8):

$$2CO = C + CO_2 \tag{8}$$

Similar to the methane cracking, a carbon activity can be defined for Boudouard reaction ($a_c^b$) by using the local concentration/partial pressure of the gases involved, as presented in Equation (9):

$$a_c^b = K_c^b \frac{(p_{CO})^2}{p_{CO_2}} \tag{9}$$

Simulated distribution of the carbon formation risks from Boudouard reaction and methane cracking are depicted in Figures 7 and 8, respectively. Figure 7 shows that under the operational conditions used in this study and fuel utilization ranging from 65% to 75%, the probability of carbon formation through Boudouard reaction is low and well below 1 regardless of the fuel reforming pattern, i.e., fast or distributed reformation patterns. This is primarily due to the fact that this reaction is not thermodynamically benefited by the elevated temperature specifically above 700 °C. The trend along the anode length can be explained with respect to the CO and $CO_2$ concentrations profiles and the anode thermal behaviour. The risk is relatively high at lower fuel utilization levels due to relatively higher and lower levels of CO and $CO_2$, respectively, in the anode gas mixture and also lower local temperature, all enhancing the Boudouard reaction chance to occur. Note that according to Equation (9), at any given temperature, the higher the ratio of $(p_{CO})^2/p_{CO2}$, the higher the carbon activity. Near the inlet, CO is low, as methane reforming has not progressed much. Further down, the ratio depends on how much CO is formed by reforming and how much $CO_2$ is formed by WGS (Equation (2)). The WGS equilibrium is also affected by the current draw, as the WGS equilibrium shifts to the right as more $H_2$ are consumed by the hydrogen electrochemical oxidation reaction (Equation (3)). It will also be affected by electrochemical oxidation of CO to $CO_2$, but the extent of this reaction is generally small, as this reaction is not as fast as $H_2$ oxidation. Moreover, due to equal stoichiometry, this effect is accounted for by the WGS equilibrium reaction (Equation (2)). As the WGS reaction generates one mole of hydrogen per mole of CO, the electrochemical and chemical balance is unaffected whether the CO conversion is modeled as a WGS or as electrochemical oxidation. In a recent work [23], the effect of different reaction kinetics and equilibrium of the methane steam reforming reaction and WGS reaction were shown to have significant effect on the concentration profiles along the cell length. Temperature change along the cell length also affects the carbon activity as it changes the value of the equilibrium constant.

Figure 8 shows a very high risk of carbon formation by methane cracking near the anode inlet where methane concentration is high. The risk increases for lower fuel utilization and for slower reforming activity. The carbon formation activity is calculated based on thermodynamic equilibrium; in practice, whether carbon will be formed will depend on the kinetics of the reactions involved.

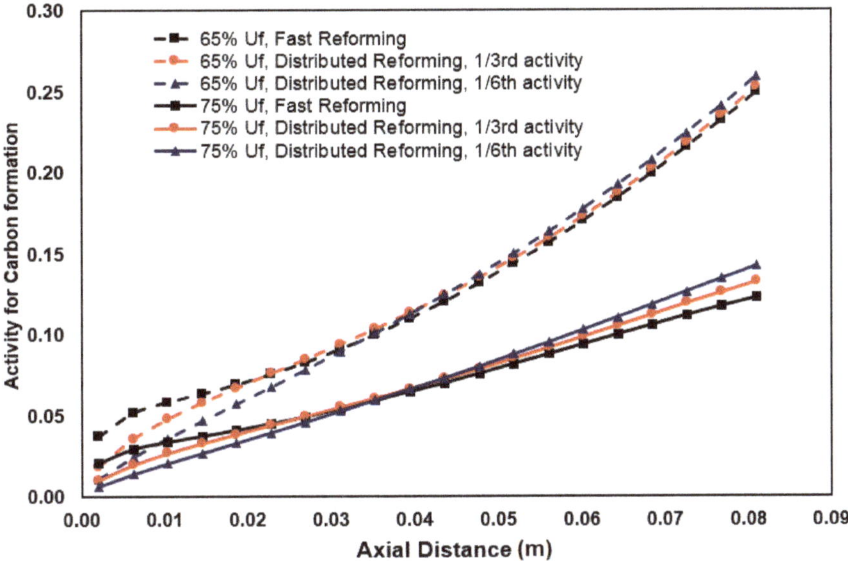


**Figure 7.** Profile of carbon formation risk from Boudouard reaction for different reforming activity at 65% and 75% fuel utilization.

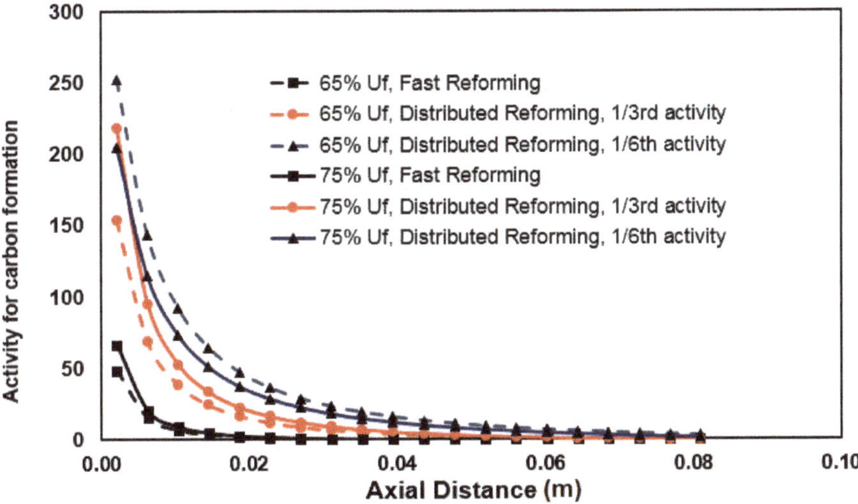


**Figure 8.** Profile of carbon formation risk from methane cracking for different reforming activity at 65% and 75% fuel utilization.

## 4. Conclusions

A model of a SOFC cell incorporating a 1D model of the anode has been developed in Aspen HYSYS. A salient feature of this model is its ability to predict simultaneous direct internal reforming on the anode and electrochemical reaction with these two reactions thermally integrated. In spite of being based on empirical correlations that makes modeling platform computationally efficient it successfully captures the distributed variables associated with stack level performance, particularly with respect to internal reforming kinetics and electrical performance. This has been achieved through interlinking of the in-built PFR module with a spreadsheet block inside the Aspen HYSYS environment. Cases for various levels of reforming activity have been compared to demonstrate the effect and relative advantages and disadvantages in terms of temperature and current density profiles. Two technically challenging aspects of SOFC operation: Anode oxidation risk and carbon formation potential have been evaluated. The methodology proposed in this paper is flexible to deploy more detailed fundamental correlations such as explicit equations based on exchange current.

**Author Contributions:** Conceptualization, K.A., A.A., and M.O.T.; methodology, K.A. and A.A.; software, K.A. and A.A.; validation, K.A. and A.A.; formal analysis, K.A., A.A., and M.O.T.; investigation, K.A., A.A., and M.O.T.; resources, M.O.T.; data curation, K.A. and A.A.; writing—original draft preparation, K.A., M.O.T., and A.A.; writing—review and editing, M.O.T., K.A. and A.A.; visualization, K.A. and A.A.; supervision, M.O.T.; project administration, M.O.T. All authors have read and agreed to the published version of the manuscript.

**Funding:** This research received no external funding.

**Conflicts of Interest:** The authors declare no conflict of interest.

## References

1. Gasik, M. *Materials for Fuel Cells*; Woodhead Publishing: Sawston, Cambridge, UK, 2008.
2. Cooper, S.J.; Brandon, N.P. Chapter 1—An Introduction to Solid Oxide Fuel Cell Materials, Technology and Applications. In *Solid Oxide Fuel Cell Lifetime and Reliability*; Brandon, N.P., Ruiz-Trejo, E., Boldrin, P., Eds.; Academic Press: Cambridge, MA, USA, 2017; pp. 1–18.
3. Wang, J.; Wang, H.; Fan, Y. Techno-Economic Challenges of Fuel Cell Commercialization. *Engineering* **2018**, *4*, 352–360. [CrossRef]
4. Ahmed, K.; Föger, K. Perspectives in Solid Oxide Fuel Cell-Based Microcombined Heat and Power Systems. *J. Electrochem. Energy Convers. Storage* **2017**, *14*, 031005. [CrossRef]
5. Lanzini, A.; Ferrero, D.; Santarelli, M. Energy System Analysis of SOFC Systems. In *Advances in Medium and High Temperature Solid Oxide Fuel Cell Technology*; Boaro, M., Salvatore, A.A., Eds.; Springer International Publishing: Cham, Switzerland, 2017; pp. 223–264.
6. Van Biert, L.; Visser, K.; Aravind, P.V. Intrinsic methane steam reforming kinetics on nickel-ceria solid oxide fuel cell anodes. *J. Power Sources* **2019**, *443*, 227261. [CrossRef]
7. Van Biert, L.; Godjevac, M.; Visser, K.; Aravind, P.V. Dynamic modelling of a direct internal reforming solid oxide fuel cell stack based on single cell experiments. *Appl. Energy* **2019**, *250*, 976–990. [CrossRef]
8. CFCL Delivers Fuel Cell Components for 40 Integrated m-CHP Units. Available online: http://www.fuelcelltoday.com/news-archive/2013/october/cfcl-delivers-fuel-cell-components-for-40-integrated-m-chp-units/ (accessed on 20 May 2019).
9. Amiri, A.; Vijay, P.; Tadé, M.O.; Ahmed, K.; Ingram, G.D.; Pareek, V.; Utikar, R. Solid oxide fuel cell reactor analysis and optimisation through a novel multi-scale modelling strategy. *Comput. Chem. Eng.* **2015**, *78*, 10–23. [CrossRef]
10. Doherty, W.; Reynolds, A.; Kennedy, D. Simulation of a tubular solid oxide fuel cell stack operating on biomass syngas using aspen plus. *J. Electrochem. Soc.* **2010**, *157*, B975–B981. [CrossRef]
11. Kupecki, J.; Badyda, K. SOFC-based micro-CHP system as an example of efficient power generation unit. *Arch. Thermodyn.* **2011**, *32*, 33. [CrossRef]
12. Riensche, E.; Meusinger, J.; Stimming, U.; Unverzagt, G. Optimization of a 200 kW SOFC cogeneration power plant. Part II: Variation of the flowsheet. *J. Power Sources* **1998**, *71*, 306–314. [CrossRef]
13. Tanim, T.; Bayless, D.J.; Trembly, J.P. Modeling a 5 kWe planar solid oxide fuel cell based system operating on JP-8 fuel and a comparison with tubular cell based system for auxiliary and mobile power applications. *J. Power Sources* **2014**, *245*, 986–997. [CrossRef]
14. Amiri, A.; Vijay, P.; Tadé, M.O.; Ahmed, K.; Ingram, G.D.; Pareek, V.; Utikar, R. Planar SOFC system modelling and simulation including a 3D stack module. *Int. J. Hydrogen Energy* **2016**, *41*, 2919–2930. [CrossRef]
15. Tang, S.; Amiri, A.; Tadé, M.O. System Level Exergy Assessment of Strategies Deployed for Solid Oxide Fuel Cell Stack Temperature Regulation and Thermal Gradient Reduction. *Ind. Eng. Chem. Res.* **2019**, *58*, 2258–2267. [CrossRef]

16. Zhang, W.; Croiset, E.; Douglas, P.L.; Fowler, M.W.; Entchev, E. Simulation of a tubular solid oxide fuel cell stack using AspenPlusTM unit operation models. *Energy Convers. Manag.* **2005**, *46*, 181–196. [CrossRef]
17. Anderson, T.; Vijay, P.; Tade, M.O. An adaptable steady state Aspen Hysys model for the methane fuelled solid oxide fuel cell. *Chem. Eng. Res. Des.* **2014**, *92*, 295–307. [CrossRef]
18. Ahmed, K.; Foger, K. Kinetics of internal steam reforming of methane on Ni/YSZ-based anodes for solid oxide fuel cells. *Catal. Today* **2000**, *63*, 479–487. [CrossRef]
19. Tingey, G.L. Kinetics of the Water—Gas Equilibrium Reaction. I. The Reaction of Carbon Dioxide with Hydrogen. *J. Phys. Chem.* **1966**, *70*, 1406–1412. [CrossRef]
20. Aguiar, P.; Chadwick, D.; Kershenbaum, L. Modelling of an indirect internal reforming solid oxide fuel cell. *Chem. Eng. Sci.* **2002**, *57*, 1665–1677. [CrossRef]
21. Ahmed, K.; Foger, K. An experimental and modelling study of the performance of a single-cell SOFC stack operating on mixtures of H2-CO-H2O-CO2. In Proceedings of the 4th European SOFC Forum, Oberrohrdorf, Switzerland, 30 June–3 July 2000; pp. 315–324.
22. Twigg, M.V. Catalyst Handbook. CRC: Boca Raton, FL, USA, 1989.
23. Ahmed, K.; Föger, K. Analysis of equilibrium and kinetic models of internal reforming on solid oxide fuel cell anodes: Effect on voltage, current and temperature distribution. *J. Power Sources* **2017**, *343*, 83–93. [CrossRef]

*Article*

# Studies on Influence of Cell Temperature in Direct Methanol Fuel Cell Operation

**R. Govindarasu [1,*] and S. Somasundaram [2]**

1 Department of Chemical Engineering, Sri Venkateswara College of Engineering, Pennalur, Sriperumbudur Tk-602 117, India
2 Department of Electronics and Communication Engineering, Alagappa Chettiar Government College of Engineering and Technology, Karaikudi-630 003, India; ssundaramau@gmail.com
* Correspondence: rgovind@svce.ac.in

Received: 2 January 2020; Accepted: 17 March 2020; Published: 19 March 2020

**Abstract:** Directmethanol fuel cells (DMFCs) offer one of the most promising alternatives for the replacement of fossil fuels. A DMFC that had an active Membrane Electrode Assembly (MEA) area of 45 $cm^2$, a squoval-shaped manifold hole design, and a Pt-Ru/C catalyst combination at the anode was taken for analysis in simulation and real-time experimentation. A mathematical model was developed using dynamic equations of a DMFC. Simulation of a DMFC model using MATLAB software was carried out to identify the most influencing process variables, namely cell temperature, methanol flow rate and methanol concentration during a DMFC operation. Simulation results were recorded and analyzed. It was observed from the results that the cell temperature was the most influencing process variable in the DMFC operation, more so than the methanol flow rate and the methanol concentration. In the DMFC, real-time experimentation was carried out at different cell temperatures to find out the optimum temperature at which maximum power density was obtained. The results obtained in simulation and the experiment were compared and it was concluded that the temperature was the most influencing process variable and 333K was the optimum operating temperature required to achieve the most productive performance in power density of the DMFC.

**Keywords:** direct methanol fuel cell; methanol crossover; power density; catalyst; membrane electrode assembly

## 1. Introduction

In the present juncture of the energy crisis, it has become inevitable to find alternatives for the mainly exploited fossil fuels. Recent developments in the field of fuel cells have given encouraging results suggesting their possible use of replacing the conventional highly polluting, less efficient combustible engines [1,2]. In this context, the direct methanol fuel cell (DMFC) appears to be the most promising tool to provide power to portable electronic devices. This is an electrochemical cell that has advantages such as offering a simple and easy method to store fuel, a simple design and green emissions. DMFC is a subcategory of a proton exchange membrane fuel cell (PEMFC) in which methanol is used as fuel. The salient features of DMFCs are the ease of transport of methanol, low or zero emissions, reliability in operation and utilization of methanol directly as a fuel to convert chemical energy into electric power. These DMFCs are designed especially for portable applications, where energy and power density are more important than efficiency [3,4]. A simple DMFC consists of a methanol distributor, gasket, anode, membrane, cathode, and oxygen gas distributor.

Regarding the working principles of the DMFC system, methanol is oxidized to hydrogen ions ($H^+$) and electrons ($e^-$) at the anode. The released electrons are transported from the anode to the cathode through an electrical circuit where power is withdrawn, and at the same time, the hydrogen ions travel to the cathode through the electrolyte membrane. At the cathode, both the electrons and hydrogen ions react with oxygen and produce water and heat [5].

Increasing demand for clean sources of energy and sustainable energy development has led to the exploration of alternative energy. One such activity was to develop a membrane-protected anode in a fuel cell [6]. This was conductedusingan anionic backbone of sulfonated polystyrene block–(ethylene–ran–butylene)–block polystyrene polymer on top of an anode which was used to increase the oxygen evolution. Another method developed to increase the enhancement of oxygen deposition was by selective oxidation using a cation-selective polymer material such as Nafion which improved the electrolysis and enhanced the oxygen evolution. The evolved oxygen can be used in a DMFC wherein electricity is generated [7,8]. In a direct methanol fuel cell, a solution of methanol is internally reformed. This is conducted with the help of a suitable catalyst, which is then oxidized at the anode and liberates electrons and protons. Currently, the Hilbert fractal curve is used to design a DMFC [9]. It is a continuous space-filling curve that can be applied to grids of power. These curves are used to study the current collectors of the direct methanol fuel cell.

Currently, nanostructured catalysts lead to enhanced efficiency, robustness, and reliability in energy conversion and storage systems due to their unique physicochemical and electrochemical properties [10]. Carbon nanofiber webs have beenused as a porous methanol barrier to reduce the effect of methanol crossover in DMFCs. They have enhanced the cell performance at low methanol concentrations due to their balanced effect on reactant and product management, with an increase in peak power density compared to the conventional DMFC [11]. A mathematical model was developed and validated in realtime, to study the transient temperature distribution across a DMFC. The model was used to study temperature distribution across passive and active DMFCs as a function of process parameters and performance parameters [12].

Advanced control strategies based on methanol concentration were designed and implemented to increase the efficiency of a DMFC. This method is more efficient than performing several modifications to system layouts and operating strategies [13]. In addition to the control of an integrated fuel processing system, the fuel cell system in a DMFC provides efficient fuel cell operation and sustainable power source for various utilities [14].

In this study, a small DMFC with a 45 $cm^2$sizedmembrane electrode assembly (MEA) was fabricated, and the effects of cell temperature on the DMFC performance were studied in simulation and in realtime.

## 2. Materials and Methods

### 2.1. Model-Based Simulation

Design of fuel cells does not necessarily mean higher efficiency. Improvement of fuel cell efficiency is much dependent on the modelling and control operations [15,16]. A Laplace domain model of a DMFC was developed using six different first-order equations which represented the DMFC operation (Table 1, Table 2). These equations were obtained from the material balances, potential balance against the anode and the cathode, coverage of species on the catalyst surface, and cell current. Aqueous methanol in the feed channel was considered to be in direct contact with the catalytic layer. Using the mathematical modelling equations of a DMFC, the MATLAB program was written with the s-function technique.

**Table 1.** Equation for the variation of $C_{CH3OH}$ and theanode/cathode overpotential.

| S.No. | Variation of $C_{CH3OH}$ and The Anode/Cathode Over Potential |
|---|---|
| 1 | $\frac{dC_{CH_3OH}}{dt} = \frac{1}{\tau}\left(C^F_{CH_3OH} - C_{CH_3OH}\right) - \frac{A_S}{V}N_{CH3OH} - \frac{A_S}{V}r1$ |
| 2 | $\frac{d\eta_A}{dt} = \frac{1}{C_a}\left(i_{Cell} - F(3r1 + 3r2)\right)$ |
| 3 | $\frac{d\eta_c}{dt} = \frac{1}{C_c}\left(-i_{Cell} - 6F\left(r5 + N_{CH_3OH}\right)\right)$ |

**Table 2.** Equation for coverage of adsorbed species at the anode.

| S.No. | Coverage of Adsorbed Species at Anode |
|---|---|
| 1 | $\frac{dPt3-COH}{dt} = \frac{1}{\gamma C_t}(r1 - r3)$ |
| 2 | $\frac{d\theta_{Ru-OH}}{dt} = \frac{1}{\gamma C_t}(3r2 - 2r3 - r4)$ |
| 3 | $\frac{dPt3-COOH}{dt} = \frac{1}{\gamma C_t}(r3 - r4)$ |

This program is used to compute the DMFC output voltage based on the derived anode overpotential, cathode overpotential, and the cell current as given below:

$$U = U_0 - \eta_A + \eta_C - \frac{d^M}{k^M} i_{Cell} \tag{1}$$

The output voltage $U$ has been influenced by three variables, namely, the flow rate of methanol, the concentration of methanol, and the cell temperature. Among these variables, it was necessary to investigate which variable had a larger influence on the output voltage [17]. The developed model was simulated using MATLAB software (Figure 1). The influence of the above three variables over the DMFC output voltage was recorded and analyzed.

## 2.2. Real-Time Experimentation

In this research work, a single DMFC system was designed with a 45 $cm^2$ active cross-sectional area (Figure 2). The cell was molded into a frame of a fiber-reinforced Teflon-coated jacket and was kept between two graphite blocks. Flow grooves for methanol and oxygen flow were provided in this system. The flow field comprised a series of 25 parallel channels with a 2 mm depth, a 1 mm wide rib and a 1 mm groove [18]. A provision for the tapping current and voltage measurement was provided. Electrical plate type heaters were placed behind each of the graphite blocks to heat the cell for achieving the desired operating temperature. Methanol solution was circulated with the peristaltic pump. Oxygen was supplied with 2 atm pressure. Catalysts of Pt-Ru/C and Pt/C were loaded at concentrations of 2 mg/$cm^2$ at anode and cathode, respectively. Nafion 117, a polymer electrolyte membrane, was used in MEA preparations. It has a high protonic conductivity, zero electronic conductivity, excellent mechanical stability, and low resistance. This polymer electrolyte membrane also acts as a separator between the anode and the cathode. A graphite plate is commonly used as reactant distributor cum current collector [19,20]. The reactions involved in the DMFC operation were as follows:

$$\text{Anode reaction}: CH_3OH + H_2O \rightarrow CO2 + 6H^+ + 6e^-$$
$$\text{Cathodereaction}: 3/2O_2 + 6H^+ + 6e^+ \rightarrow 3H_2O$$
$$\text{Overallreaction}: CH_3OH + 3/2O_2 \rightarrow CO_2 + 2H_2O$$

**Figure 1.** Real-time experimental setup of the direct-methanol fuel cell (DMFC) system.

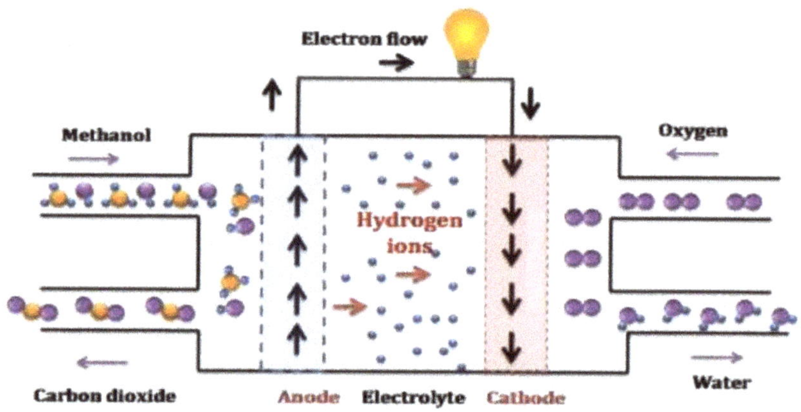


**Figure 2.** Schematic diagram of the direct methanol fuel cell system.

## 3. Results and Discussion

### 3.1. Model-Based Simulation

A program written in MATLAB with the mathematical modelling equations of the DMFC was simulated as discussed in Section 2.1. In order to find out the most influencing variable of the DMFC, the analysis was carried out for cell temperature, methanol flow rate, and methanol concentration. During the simulation, the effect of different operating temperatures between 303 K and 343 K, different methanol flow rates in the ranges of 0.25 CCM to 2 CCM and different methanol concentrations in the ranges of 0.25 M to 2 M were recorded. The results obtained are plotted in Figures 3–5.

It was also observed from Figures 3 and 4 that the DMFC output voltage increased the methanol flow rate by up to 1 CCM and increased the 1M methanol concentration, thereafter, (i.e., greater than 1CCM methanol flow rate and higher than 1M methanol concentration) DMFC output voltage did not vary significantly and remained constant due to the methanol crossover to cathode. It was also observed from Figure 5 that the output voltage of the DMFC increased linearly with an increase in temperature up to 333 K and decreased beyond that due to the dry-out effect in the cell. From the dynamic studies, it is clear that the temperature is the most influencing variable on the output voltage of DMFC. Hence, this working temperature was selected as a manipulated variable for further investigation.

Step response analysis was carried out in simulation to validate the developed mathematical model and its behavior. Step input changes were given at the operating temperatures of 313 K, 323 K, and 333 K (25%, 50%, 75% of the cell temperature) with ± 10% and ± 15%. Step response of the system at 25% of the cell temperature for a step change of ± 10%, and a step change of ± 15% of cell temperature were observed and are given in Figure 6. Similar runswere conducted at 50% of the cell temperature and 75% of cell temperature and are recorded in Figures 7 and 8, respectively. From the step response curves, it was concluded that the developed model was behaving linearly in these operating temperatures.

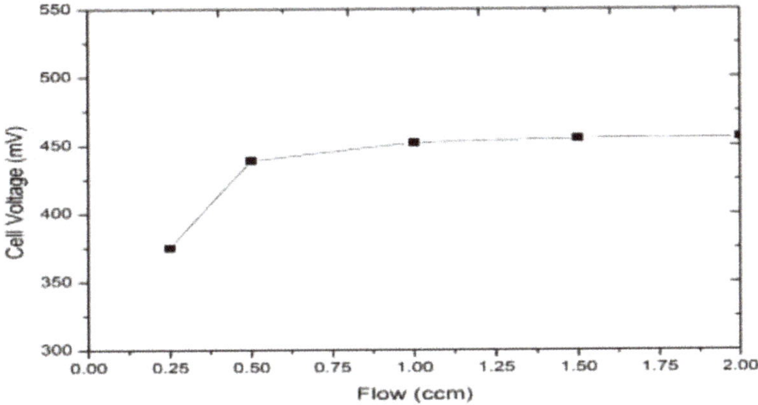


**Figure 3.** Effect of methanol flowrate.

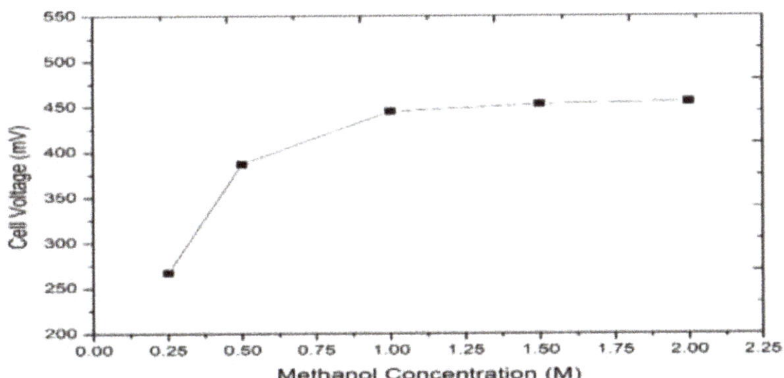


**Figure 4.** Effect of methanol concentration.

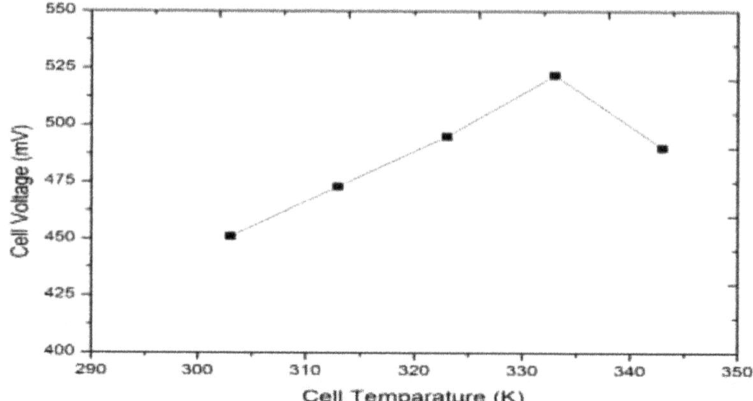


**Figure 5.** Effect of cell temperature.

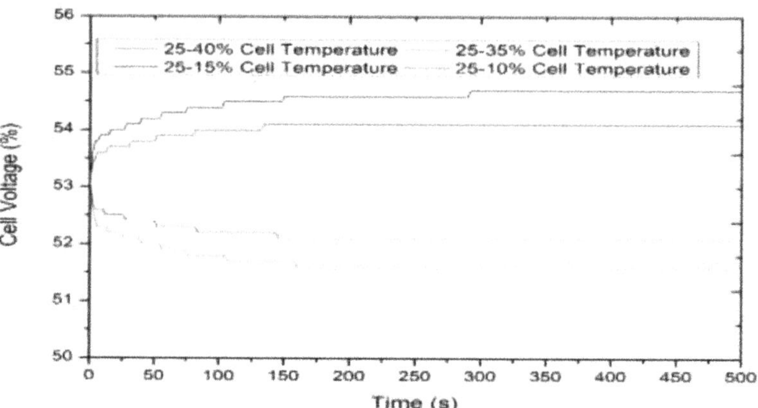


**Figure 6.** Step response of DMFC at 25% ofthe operating temperature.

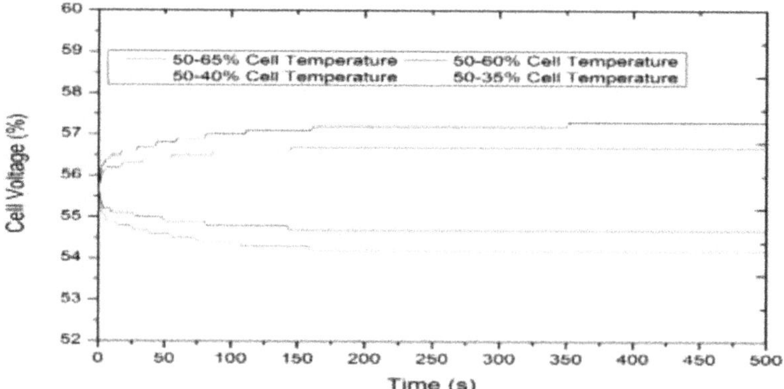


**Figure 7.** Step response of DMFC at 50% of the operating temperature.

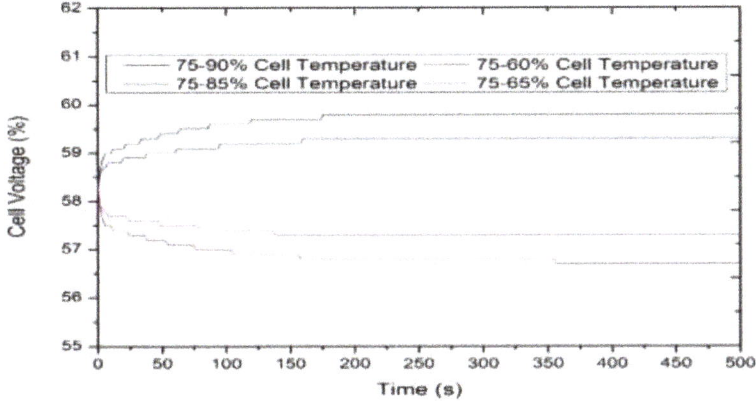


**Figure 8.** Step response of DMFC at 75% of the operating temperature.

## 3.2. Real-Time Experimentation

Experimentation of the squoval-shaped manifold holedesign(SSMHD)-based DMFC was carried out with a Pt-Ru/C catalyst combination in order to determine the optimum operating temperature ($\mathbf{T}_{\mathbf{Cell}}$) of the DMFC with a methanol concentration of 1 M and a methanol flow rate of 1 mL/min. The output voltage of DMFC for a defined current was measured using an electronic load bank and the performance of DMFC was analyzed. Voltage and power density obtained during the run of the experiment against current density were plotted (I–V Curve and I–P Curve) and are shown in Figures 9 and 10, respectively.

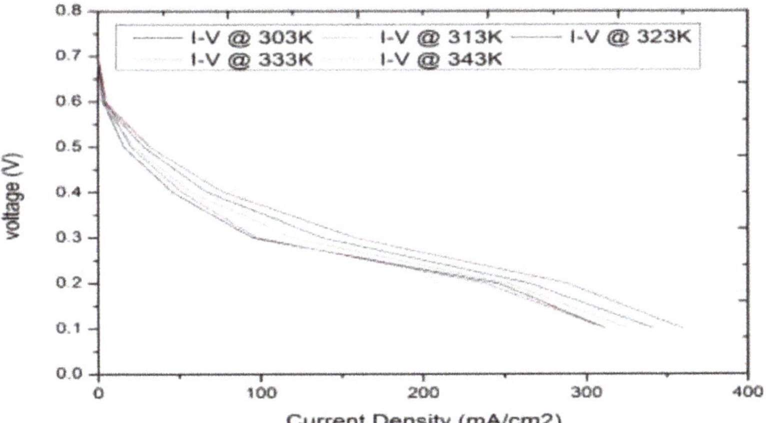


**Figure 9.** $C_{CH3OM}$: Steady-state current density–voltage curve.

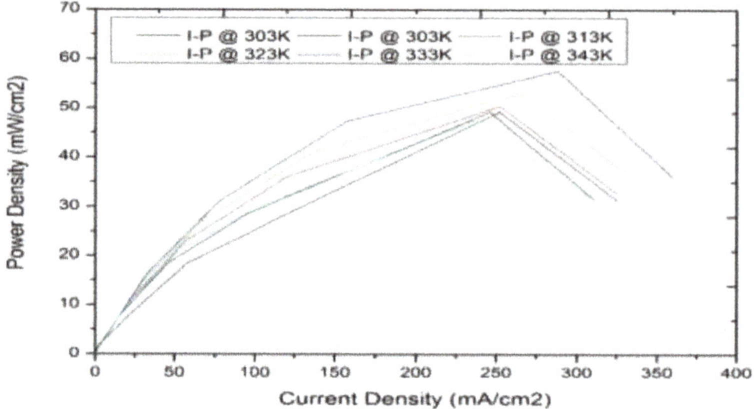


**Figure 10.** $C_{CH3OM}$: Steady-state current density–power density curve.

It was observed from Figures 9 and 10 that an increase in the DMFC temperature from 303 K (30°C) to 333 K (60°C) led to an increase in DMFC performance, butfor a further increase in temperature beyond its boiling point 333 K (60°C), DMFC performance was reduced due to the dry-out effect. Hence, it is reported that 333 K (60°C) is the optimum operating temperature at which a maximum power density of 57.6 mW/cm$^2$ was obtained for a squoval-shaped manifold design (SSMD)-incorporated direct methanol fuel cell. In addition to the studies on temperature, the durability of a DMFC at different load currents, namely 2A, 4A, 6A, and 10A, was also studied.Output voltages were recorded and are shown in Figure 11. It was confirmed by the consistent, steady output that the SSMD-based DMFC was durable at the optimized operating conditions, namely a DMFC temperature of 333 K, a methanol concentration of 1 M, and a methanol flow rate of 1 mL/min.

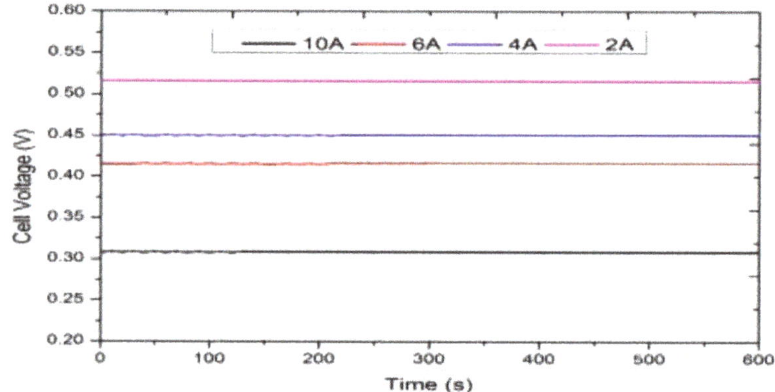


**Figure 11.** Durability test on a squoval-shaped manifold hole design-based DMFC.

## 4. Conclusions

A direct methanol fuel cell (DMFC) is a particular type of fuel cell that has its own merits. In this work, a direct methanol fuel cell with a 45 cm$^2$ activation area in the MEA, and a Pt-Ru/C catalyst at the anode was fabricated and tested with the squoval-shaped manifold hole design-incorporated DMFC work station available in the laboratory. Performance of the DMFC was evaluated with different operating parameters, namely cell temperature, methanol flow rate, and methanol concentration in simulation. From the simulation results, it was concluded that cell temperature was the most influencing process variable. From the results, the optimum temperature ($T_{Cell}$) of the DMFC operation was identified as 333 K. Real-time experimentation was carried out with different cell temperatures, and the results were recorded. It was observed from the experimental data that the maximum power density of 57.6 mW/cm$^2$ at 288 mA/cm$^2$ was achieved with the said operating temperature of 333 K. With this operating temperature, the durability of the DMFC was also verified at different load currents and found to be durable.

**Author Contributions:** This paper is come to the final shape with the conceptualization of R. Govindarasu and recourse of S. Somasundaram. All authors have read and agreed to the published version of the manuscript.

**Funding:** This research received no external funding.

**Acknowledgments:** The authors are grateful to P. Kanthabhaba, Former Professor and Head, Department of Chemical Engineering, Annamalai University, Chidambaram-608002, India.

**Conflicts of Interest:** The authors declare no conflict of interest.

## References

1. Dinesh, J.; Easwaramoorthi, M.; Muthukumar, M. State of Research Developments in Direct Methanol Fuel cell. *Int. J. Eng. Trends Technol.* **2017**, *43*, 284–296. [CrossRef]
2. Dillon, R.; Srinivasan, S.; Arico, S.A.; Antonucci, V. International activities in DMFCR&D: Status of technologies and potential applications. *J. Power Sources* **2004**, *127*, 112–126.
3. Vielstich, W.; Gasteiger, H.; Lamm, A. *Handbook of Fuel Cell: Fundamentals, Technology, Applications*; John Wiley & Sons: New York, NY, USA, 2003.
4. Arico, S.; Srinivasan, S.; Antonucci, V. DMFCs: From Fundamental Aspects to Technology Development. *Fuel Cells* **2001**, *1*, 133–161. [CrossRef]
5. Bostaph, J.; Xie, C.G.; Pavio, J.; Fisher, A.M.; Mylan, B.; Hallmark, J. Design of a 1-W direct methanol fuel cell system in Direct Methanol Fuel Cell. In Proceedings of the 40 th Power Sources Conference, Cherry Hill, NJ, USA, 10–13 June 2002; pp. 211–214.
6. Venkatkarthick, R.; Elamathi, S.; Sangeetha, D.; Balaji, R.; Kannan, B.S.; Vasudevan, S.; Davidson, D.J.; Sozhan, G.; Ravichandran, S. Studies on polymer modified metaloxideanodeforoxy gen evolution reaction in saline water. *J. Electroanal. Chem.* **2013**, *697*, 1–4. [CrossRef]
7. Balaji, R.; Kannana, B.S.; Lakshmi, J.; Senthil, N.; Vasudevan, S.; Sozhan, G.; Shukla, A.K.; Ravichandran, S. An alternative approach to selective seawater oxidation for hydrogen production. *Electrochem. Commun.* **2009**, *11*, 1700–1702. [CrossRef]
8. Wang, C.Y. Principles of direct methanol fuel cells for portable and micropower. In *Mini-Micro Fuel Cells;* Springer: Ankara, Turkey, 2008; pp. 235–242.
9. Chang, J.Y.; Kuan, Y.D.; Lee, S.M. Experimental Investigation of a Direct Methanol Fuel Cell with Hilbert Fractal Current Collectors. *Hindawi-J. Chem.* **2014**, *2014*, 371616. [CrossRef]
10. Balasinga, S.K.; Nallathamb, K.S.; Jabbar, M.H.A.; Ramadoss, A.; Kamaraj, S.K.; Nanomaterials, M.K. Electrochemical Energy Conversion and Storage Technologies. *Hindawi-J. Nanomater.* **2019**. [CrossRef]
11. Yuan, W.; Fang, G.; Li, Z.; Chen, Y.; Tang, Y. Using Electrospinning-Based Carbon Nanofiber Webs for Methanol Crossover Control in Passive Direct Methanol Fuel Cells. *Materials* **2018**, *11*, 71. [CrossRef] [PubMed]
12. Ramesh, V.; Krishnamurthy, B. Modeling the transient temperature distribution in a Direct Methanol fuel cell. *J. Electroanal. Chem.* **2018**, *809*, 1–7. [CrossRef]
13. Youngseung, N.; Zenith, F.; Krewer, U. Increasing Fuel Efficiency of Direct Methanol Fuel Cell Systems with Feed forward Control of the Operating Concentration. *Energies* **2015**, *8*, 10409–10429. [CrossRef]
14. Ohenoj, M.; Ruusunen, M.; Leivisk, K. Hierarchical Control of an Integrated Fuel Processing and Fuel Cell System. *Materials* **2019**, *12*, 21. [CrossRef] [PubMed]
15. Simoglou, A.; Argyropoulos, P.; Martin, E.B.; Scott, K.; Morris, A.J.; Taama, W.M. Dynamics modeling of the voltage response of direct methanol fuel cell sandstacks Part I: Mode development and validation. *Chem. Eng. Sci.* **2001**, *56*, 6761–6772. [CrossRef]
16. Govindarasu, R.; Parthiban, R.; Bhaba, P.K. Recent evolutions in modelling of Direct Methanol Fuel Cell. *Elixir Int. J. Chem. Eng.* **2014**, *72*, 25428–25433.

17. Maynard, H.L.; Meyers, J.P. Miniature fuel cells for portable power: Design considerations and challenges. *J. Vac. Sci. Technol.* **2002**, *20*, 1287–1297. [CrossRef]
18. Govindarasu, R.; Parthiban, R.; Bhaba, P.K. Investigation of Flow Maldistribution in proton exchange membrane fuel cell. *Int. J. Renew. Energy Res.* **2012**, *2*, 653–656.
19. Manokaran, A.; Vijayakumar, R.; Sridhar, P.; Pitchumani, S.; Shukla, K. A self-supported Direct methanol fuel cell system. *J. Chem. Sci.* **2011**, *123*, 343–347. [CrossRef]
20. Premkumar, S.; Prabhakar, S.; Lingeswaran, K.; Ramnathan, P. Development of Direct Methanol Fuel Cell and Improving the efficiency. *Middle-East J. Sci. Res.* **2014**, *20*, 1277–1280.

Article

# Air-Forced Flow in Proton Exchange Membrane Fuel Cells: Calculation of Fan-Induced Friction in Open-Cathode Conduits with Virtual Roughness

**Dejan Brkić [1,2,*,†] and Pavel Praks [1,*,†]**

1 IT4Innovations, VŠB—Technical University of Ostrava, 708 00 Ostrava, Czech Republic
2 Faculty of Electronic Engineering, University of Niš, 18000 Niš, Serbia
* Correspondence: dejanbrkic0611@gmail.com or dejan.brkic@vsb.cz or dejan.brkic@elfak.ni.ac.rs (D.B.); pavel.praks@vsb.cz (P.P.)
† Both authors contributed equally to this article.

Received: 8 April 2020; Accepted: 3 June 2020; Published: 11 June 2020

**Abstract:** Measurements of pressure drop during experiments with fan-induced air flow in the open-cathode proton exchange membrane fuel cells (PEMFCs) show that flow friction in its open-cathode side follows logarithmic law similar to Colebrook's model for flow through pipes. The stable symbolic regression model for both laminar and turbulent flow presented in this article correlates air flow and pressure drop as a function of the variable flow friction factor which further depends on the Reynolds number and the virtual roughness. To follow the measured data, virtual inner roughness related to the mesh of conduits of fuel cell used in the mentioned experiment is 0.03086, whereas for pipes, real physical roughness of their inner pipe surface goes practically from 0 to 0.05. Numerical experiments indicate that the novel approximation of the Wright-ω function reduced the computational time from half of a minute to fragments of a second. The relative error of the estimated friction flow factor is less than 0.5%.

**Keywords:** Colebrook equation; fuel cells; flow friction factor; open-cathode; pressure drop; symbolic regression; numerically stabile solution; roughness

---

## 1. Introduction

Flow friction is a complicated physical phenomenon and it is not constant but depends on flow rate and pressure drop. Because of its complexity, equations which describe the flow friction are mostly empirical [1]. The most used empirical formulation for turbulent pipe flow is given by Colebrook's equation [2]. Here we will evaluate flow friction factor caused by air flow in the cathodic side of the observed PEMFCsand we will provide accurate and consistent solution.

### 1.1. Colebrook Equation for Pipe Flow Friction

The standard Colebrook's friction equation for turbulent pipe flow [2]; Equation (1), follows logarithmic law and is based on an experiment performed by Colebrook and White in the 1930s [3], while its graphical interpretation was given by Moody in 1944 [4].

$$\frac{1}{\sqrt{\lambda_{T(p)}}} = -2 \cdot log_{10}\left(\frac{2.51}{Re_{(p)}} \cdot \frac{1}{\sqrt{\lambda_{T(p)}}} + \frac{\varepsilon_{(p)}}{3.71}\right) \quad (1)$$

where:

$\lambda_{T(p)}$—turbulent Darcy flow friction factor for pipes (dimensionless)

$Re_{(p)}$—Reynolds number (dimensionless)—the same definition as for fuel cells
$\varepsilon_{(p)}$—relative roughness of inner pipe surface (dimensionless)
$log_{10}$—logarithmic function with base 10
$p$—index related to pipes

In Moody's diagram, for the turbulent regime the Reynolds number $Re_{(p)}$ goes from around 2320 to $10^8$ while the relative roughness of inner pipe surface $\varepsilon_{(p)}$ from 0 for smooth surfaces to about 0.05 for very rough surfaces (on the other hand, flow friction for laminar regime for pipe flow $\lambda_{L(p)}$ does not depend of roughness $\varepsilon_{(p)}$ while the Reynolds number $Re_{(p)}$ goes from 0 to around 2320; the formula for laminar Darcy flow friction factor for pipe flow, $\lambda_{L(p)} = 64/Re_{(p)}$ is not empirical but theoretically founded).

Colebrook and White experimented with airflow through pipes of different roughness of inner surface [3]. They used a set of pipes from which one left with a smooth inner surface, while others were covered with sand of different size of grains. For each pipe, one uniform grain size with glue as adhesive material was used. Thus, the pipes in the experiment were gradually smooth to the fully rough. The experiment revealed that the turbulent flow friction depends on the Reynolds number $Re_{(p)}$ and on the relative roughness of inner pipe surface $\varepsilon_{(p)}$. As can be seen from the Moody diagram [4], for the same values of the Reynolds number $Re_{(p)}$ the turbulent flow friction factor $\lambda_{T(p)}$ is higher in the pipes with higher relative roughness $\varepsilon_{(p)}$, where subsequently for the same flow, the corresponding pressure drop is higher, too.

The Colebrook equation is implicitly given in respect to the unknown turbulent flow friction factor $\lambda_{T(p)}$ and it can be rearranged in explicit form only in terms of Lambert W-function [5] or its cognate Wright $\omega$-function, and even then further evaluation can be only approximated. Very accurate approximate formulas for the Colebrook equation for pipe flow based on the Wright $\omega$-function are available [6,7].

### 1.2. Modified Colebrook Equation for Flow Friction

The Colebrook equation is empirical [3] and therefore possible modifications based on different conditions of flow can be done. We will evaluate here a modification for fuel cells [8,9]. Further, for example, US Bureau of Mines published a report in 1956 [10] that introduced a modified form of the Colebrook equation for gas flow where coefficient 2.51 was replaced with 2.825. Also, very recent modifications of a variety of empirical equations for pipe flow are available [11]. Here we analyze one modification for air flow friction in the open-cathode side of the observed type of PEMFCs [8], and subsequently, we give a stable and computationally efficient explicit solution which is valid in this case. Analogous analyses can be done for air flow for cooling of electronic products from hand-held devices to supercomputers [12,13]. Here we discuss the only influence of hydraulic effects of flow through the open-cathode conduits of fuel cells while available literature [14–21] should be consulted for thermodynamic aspects.

The Colebrook equation can be rearranged following data obtained from the experiment by Barreras et al. with fuel cells [8]. Based on this data and following analogy with pipe flow, it is estimated that virtual roughness is fixed by value $\varepsilon_{(FC)} = 0.03086$. Therefore, the Colebrook equation can be rewritten for the observed fuel cell as Equation (2). We will further analyze this equation to provide its stable numerical solution. The value of virtual roughness $\varepsilon_{(FC)} = 0.03086$ is from [9].

$$\frac{1}{\sqrt{\lambda_{T(FC)}}} = -10 \cdot log_{10}\left(\frac{65.6}{Re_{(FC)}} \cdot \frac{1}{\sqrt{\lambda_{T(FC)}}} + \frac{\varepsilon_{(FC)}}{0.1415596}\right) \tag{2}$$

where:

$\lambda_{T(FC)}$—turbulent Darcy flow friction factor for fuel cells (dimensionless)
$Re_{(FC)}$—Reynolds number (dimensionless)—the same definition as for pipes

$\varepsilon_{(FC)}$—virtual relative roughness of fuel cell (dimensionless)
$log_{10}$—logarithmic function with base 10
$FC$—index related to Fuel Cells

## 2. Proposed Model

Proton exchange membrane fuel cells (PEMFCs) transform chemical energy from electrochemical reaction of hydrogen and oxygen to electrical energy [22–25]. Here we analyze fan-induced air-forced flow, based on data from the experiment with pressure drop in the cathode side of air-forced open-cathode PEMFCs by Barreras et al. [8]. Their experiment with fuel cells can be compared with the experiment performed by Colebrook and White with air flow through pipes [3]. Barreras et al. [8] use three different cathode configurations with aspect ratios $h/H$ from 0.83 to 2.5 to form a mesh of cathodic channels to supply the fuel cell with enough air for cooling and with oxygen for a chemical reaction (roughness is real physical characteristic of pipe surface [26], but not of cathodic channels of fuel cells in terms of hydraulic performances).

Value of the Reynolds number $Re_{(FC)}$ during the experiment was from 45 to 4000, while as a difference from pipes, during flow through the observed fuel cell, the transition from laminar to turbulent flow occurred around $Re_{(FC)} = 500$. In our numerical experiments, we use $Re_{(FC)}$ between 50 and 4100.

As already mentioned, the original Colebrook equation is valid for turbulent flow of air, water, or natural gas through pipes. On the other hand, for laminar flow, $\lambda_{L(p)} = 64/Re_{(p)}$ is used whereas the transition from laminar to turbulent flow is around $Re_{(p)} \approx 2320$. This transitional border at the Moody's plot [4] is sharp where the equivalent sharp transition from laminar to turbulent flow for the observed fuel cell starts at around $Re_{(FC)} \approx 500$, as explained in [8]. Therefore, for airflow through the cathode side of the observed fuel cells, the flow friction factor $\lambda_{L(FC)}$ consists of two clearly defined types of flow:

1. laminar flow $\lambda_{L(FC)}$ that depends both on the Reynolds number $Re_{(FC)}$ and on the geometry of conduits; height $h$ and width $H$ of the mesh of conduits that forms a mesh of cathodic air channels, and
2. turbulent flow $\lambda_{T(FC)}$ is solely the function of the Reynolds number $Re_{(FC)}$ for the case from the experiment of Barreras et al. [8] (in general also on virtual roughness [9], which is in this case $\varepsilon_{(FC)} = 0.03086$).

For solving implicitly given equations, instead of iterative procedures [27,28], appropriate explicit approximations which are accurate but also computationally efficient can be used (review of appropriate explicit approximations for pipe flow friction is available in [29]). A computationally efficient and stabile unified equation for the observed type of fuel cells which is valid both for laminar and turbulent regime will be given in this article [30].

### 2.1. Turbulent Flow

In case of turbulent airflow during experiments with open-cathode PEMFCs, measurements show that pressure drop during turbulent flow at its cathode side follows logarithmic law, which form is comparable to the Colebrook's flow friction equation for pipe flow, but with different numerical values [8]. The flow friction related to air flow is given by Equation (3):

$$\frac{1}{\sqrt{\lambda_{T(FC)}}} = -10 \cdot log_{10}\left(\frac{65.6}{Re_{(FC)}} \cdot \frac{1}{\sqrt{\lambda_{T(FC)}}} + 0.218\right) \quad (3)$$

where:
$\lambda_{T(FC)}$—turbulent Darcy flow friction factor for fuel cells (dimensionless)

$Re_{(FC)}$—Reynolds number (dimensionless) – the same definition as for pipes
$log_{10}$—logarithmic function with base 10
$FC$—index related to Fuel Cells

During turbulent flow, numerical values for the flow friction factor in pipe and fuel cells are different and that difference can go up to 60%. To make a direct connection between Equation (1) for pipe flow and Equation (3) for the observed fuel cell, Equation (4) can be used [25]:

$$\frac{1}{\sqrt{\lambda_{T(FC)}}} = -14.17 + 5 \cdot \left( -2 \cdot log_{10} \left( \frac{2.51}{Re_{(FC)}} \cdot \frac{1}{\sqrt{\lambda_{T(FC)}}} + \frac{\varepsilon_{(FC)}}{3.71} \right) \right) \quad (4)$$

where:

$\lambda_{T(FC)}$—turbulent Darcy flow friction factor for fuel cells (dimensionless)
$Re_{(FC)}$—Reynolds number (dimensionless)—the same definition as for pipes
$\varepsilon_{(FC)}$—virtual relative roughness of fuel cell (dimensionless)
$log_{10}$—logarithmic function with base 10
$FC$—index related to Fuel Cells

For Equation (4), virtual roughness can be recalculated based on the Colebrook equation as $\varepsilon_{(FC)} = 0.03086$ for the observed fuel cell in the experiment [8]. This fuel cell was tested with three different cathode configurations [8]. As noted in [31], this roughness $\varepsilon_{(FC)}$ is not a real measurable physical characteristic of the surface of the used material for conduits (on the contrary for pipes $\varepsilon_{(p)}$ can be measured or at least estimated accurately [32–37]).

Both, Equations (3) and (4) are numerically unstable for $Re_{(FC)} < 575$, which can be a critical problem knowing that turbulent zone starts for $Re_{(FC)} > 500$. However, the novel solution proposed in this article is numerically stable.

Generally, implicitly given equations can be transformed in explicit form through the Lambert W-function [38,39]. The Lambert W-function [5] is defined as the multivalued function W that satisfies $z = e^{W(z)} \cdot W(z)$. However, such transformation for the Colebrook equation for pipes contains a fast-growing term $e^x$ and because of that, overflow error in computers is possible [40,41]. Happily, results with fuel cells show that the solution is not in the zone where $e^x$ is too big to be stored in registers of computers. The model for fuel cells is given in Equation (5), while the related model for pipe flow friction model can be seen in [42].

$$\left. \begin{array}{c} \frac{1}{\sqrt{\lambda_{T(FC)}}} = a_{(FC)} \cdot W\left(e^{x_{(FC)}}\right) - \frac{Re_{(FC)}}{b_{(FC)}} \cdot \frac{\varepsilon_{(FC)}}{c_{(FC)}} \\ x_{(FC)} = ln\left(\frac{Re_{(FC)}}{a_{(FC)} \cdot b_{(FC)}}\right) + \frac{Re_{(FC)}}{a_{(FC)} \cdot b_{(FC)}} \cdot \frac{\varepsilon_{(FC)}}{c_{(FC)}} \\ a_{(FC)} = \frac{10}{ln(10)} \\ b_{(FC)} = 65.6 \\ \frac{\varepsilon_{(FC)}}{c_{(FC)}} = 0.218 \rightarrow c_{(FC)} = 0.1415596 \end{array} \right\} \quad (5)$$

where:

$\lambda_{T(FC)}$—turbulent Darcy flow friction factor for fuel cells (dimensionless)
$Re_{(FC)}$—Reynolds number (dimensionless)—the same definition as for pipes
$\varepsilon_{(FC)}$—virtual relative roughness of fuel cell (dimensionless)
$a_{(FC)}, b_{(FC)}, c_{(FC)}$—constants
$x_{(FC)}$—variable
$log_{10}$—logarithmic function with base 10
$ln$—natural logarithm
$e$—exponential function
$W$—Lambert function

$FC$—index related to Fuel Cells

The parameter $c_{(FC)}$ for fuel cell is $c_{(FC)} = 0.1415596$.

After procedures from [6,7,43], the following form for fuel cells expressed through the Lambert W-function and its cognate Wright $\omega$-function is given in Equation (6):

$$\left.\begin{array}{r}\frac{1}{\sqrt{\lambda_{T(FC)}}} = \frac{10}{ln(10)}\cdot\left[ln\left(\Delta_{(FC)}\right) + W\left(e^{x_{(FC)}}\right) - x_{(FC)}\right] \\ \frac{1}{\sqrt{\lambda_{T(FC)}}} = \frac{1}{a_{(FC)}}\cdot\left[B_{(FC)} + \omega\left(x_{(FC)}\right) - x_{(FC)}\right] \\ x_{(FC)} = B_{(FC)} + 0.218\cdot\Delta_{(FC)} \\ B_{(FC)} = ln\left(\Delta_{(FC)}\right) = ln\left(Re_{(FC)}\right) - 5.652138 \\ \Delta_{(FC)} = \frac{Re_{(FC)}\cdot a_{(FC)}}{65.6} = \frac{Re_{(FC)}}{284.9} \\ \frac{1}{a_{(FC)}} = \frac{10}{ln(10)} \approx 4.343\end{array}\right\} \quad (6)$$

where:

$\lambda_{T(FC)}$—turbulent Darcy flow friction factor for fuel cells (dimensionless)
$Re_{(FC)}$—Reynolds number (dimensionless)—the same definition as for pipes
$a_{(FC)}$—constant
$x_{(FC)}$, $\Delta_{(FC)}$, $B_{(FC)}$—variable
$ln$—natural logarithm
$W$—Lambert function
$\omega$—Wright function
$FC$—index related to Fuel Cells

However, symbolic regression applied on the explicit formulation, Equation (6), which involves $W\left(e^{x_{(p)}}\right) - x_{(p)} = \omega\left(x_{(p)}\right) - x_{(p)}$ gives very simple, but still accurate results in case of pipe flow [6,7] ([44,45] confirm these results), but unfortunately these analytical formulas, which are optimized for pipes, cannot be directly applied on the fuel cell equation. Fortunately, symbolic regression gives also very promising results for fuel cells as given in Equation (7):

$$W\left(e^{x_{(FC)}}\right) - x_{(FC)} = \omega\left(x_{(FC)}\right) - x_{(FC)} \approx \frac{26.723}{ln\left(\frac{Re_{(FC)}}{284.9}\right) + 0.218\cdot\frac{Re_{(FC)}}{284.9} + 6.2611} - 3.6795 \quad (7)$$

where:

$Re_{(FC)}$—Reynolds number (dimensionless)—the same definition as for pipes
$x_{(FC)}$—variable
$ln$—natural logarithm
$W$—Lambert function
$\omega$—Wright function
$FC$—index related to Fuel Cells

To avoid repetitive computations, parameters $\Delta_{(FC)}$ and $B_{(FC)}$ are introduced, in Equation (8). Both symbolic regression analyses were performed in Eureqa, a commercial software tool, which automates the process of model building and interpretation [46,47].

$$\left.\begin{array}{r}\frac{1}{\sqrt{\lambda_{T(FC)}}} = 4.343\cdot\left(B_{(FC)} + \frac{26.723}{B_{(FC)} + 0.218\cdot\Delta_{(FC)} + 6.2611} - 3.6795\right) \\ \Delta_{(FC)} = \frac{Re_{(FC)}}{284.9} \\ B_{(FC)} = ln\left(\Delta_{(FC)}\right)\end{array}\right\} \quad (8)$$

where:

$\lambda_{T(FC)}$—turbulent Darcy flow friction factor for fuel cells (dimensionless)
$Re_{(FC)}$—Reynolds number (dimensionless)—the same definition as for pipes
$\Delta_{(FC)}$, $B_{(FC)}$—variable

*ln*—natural logarithm
*FC*—index related to Fuel Cells

### 2.2. Unified Model

Although the expression for laminar flow through pipes is $\lambda_{L(p)} = 64/Re_{(p)}$, for fuel cells it is different, as given in Equation (9) [8]:

$$\lambda_{L(FC)} = \frac{58.91 + 50.66 \cdot e^{-\frac{3.4 \cdot h}{H}}}{Re_{(FC)}} \quad (9)$$

where:
$\lambda_{L(FC)}$—laminar Darcy flow friction factor for fuel cells (dimensionless)
$\frac{h}{H}$—channel depth/channel width used only in laminar flow (dimensionless)
*e*—exponential function
*FC*—index related to Fuel Cells
Values of $h/H$ are from 0.83 to 2.5.

The experiment [8] shows that air flow through the cathode side of air-forced open-cathode PEMFCs are (1) laminar for the lower values of the Reynolds number, $Re_{(FC)} < 500$ and (2) turbulent for the higher values, $500 < Re_{(FC)} < 4000$, where the Reynolds number is in hydraulics a very well-known dimensionless parameter that is used as a criterion for foreseeing flow patterns in a fluid's behavior (defined in the same way for air flow through pipes and here discussed air flow through fuel cells). The dimensionless Darcy's unified flow friction factor $\lambda_{(FC)}$, is the function of the switching function $y$, the laminar flow friction $\lambda_{L(FC)}$, and the turbulent flow friction $\lambda_{T(FC)}$. The unified coherent flow friction model that covers both laminar and turbulent zones is set by Equation (10) [30]:

$$\lambda_{(FC)} = y \cdot \lambda_{T(FC)} + (1 - y) \cdot \lambda_{L(FC)} \quad (10)$$

where:
$\lambda_{(FC)}$—unified Darcy flow friction factor for fuel cells (dimensionless)
$\lambda_{T(FC)}$—turbulent Darcy flow friction factor for fuel cells (dimensionless)
$\lambda_{L(FC)}$—laminar Darcy flow friction factor for fuel cells (dimensionless)
*y*—switching function
*FC*—index related to Fuel Cells
The novel switching function $y$ is given in Equation (11):

$$y = \frac{Re_{(FC)}}{Re_{(FC)} + e^{558 - Re_{(FC)}}} \quad (11)$$

where:
$Re_{(FC)}$—Reynolds number (dimensionless)—the same definition as for pipes
*y*—switching function
*e*—exponential function
*FC*—index related to Fuel Cells
The switching function was obtained by symbolic regression using HeuristicLab [47] and it is given in Figure 1.

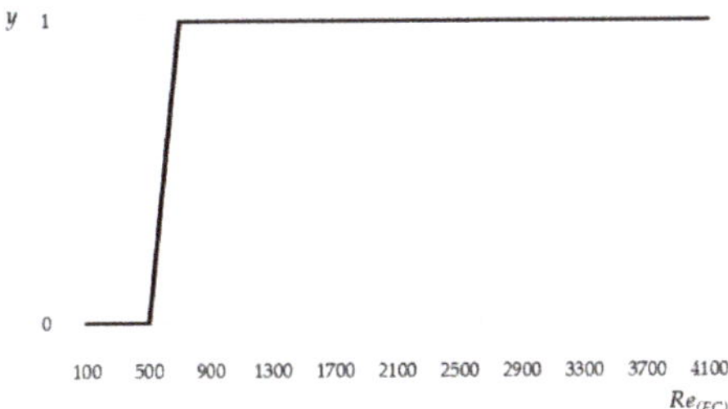


**Figure 1.** Switching function given in Equation (11).

The laminar flow friction $\lambda_{L(FC)}$ depends on the Reynolds number $Re_{(FC)}$, but also on geometry of conduits, while the turbulent flow friction $\lambda_{T(FC)}$ depends only on the Reynolds number $Re_{(FC)}$. In the case of fuel cells, both coefficients are empirical. In addition, the switching function $y$ contains the exponential function, (the similar situation is for calculation of $\lambda_{L(FC)}$ as already explained).

To avoid numerical instability, it is recommended to use the explicit approximation which gives the following unified formula in Equation (12).

$$\left.\begin{aligned} \lambda_{(FC)} &= y \cdot \lambda_{T(FC)} + (1-y) \cdot \lambda_{L(FC)} \\ \lambda_{L(FC)} &= \frac{58.91 + 50.66 \cdot e^{-\frac{3.4 \cdot h}{H}}}{Re_{(FC)}} \\ \frac{1}{\sqrt{\lambda_{T(FC)}}} &= 4.343 \cdot \left( B_{(FC)} + \frac{26.723}{x_{(FC)} + 6.2611} - 3.6795 \right) \\ \Delta_{(FC)} &= \frac{Re_{(FC)}}{284.9} \\ B_{(FC)} &= ln\left(\Delta_{(FC)}\right) \\ x_{(FC)} &= B_{(FC)} + 0.218 \cdot \Delta_{(FC)} \\ y &= \frac{Re_{(FC)}}{Re_{(FC)} + e^{558 - Re_{(FC)}}} \end{aligned}\right\} \quad (12)$$

where:

$\lambda_{(FC)}$—unified Darcy flow friction factor for fuel cells (dimensionless)
$\lambda_{T(FC)}$—turbulent Darcy flow friction factor for fuel cells (dimensionless)
$\lambda_{L(FC)}$—laminar Darcy flow friction factor for fuel cells (dimensionless)
$Re_{(FC)}$—Reynolds number (dimensionless)—the same definition as for pipes
$\frac{h}{H}$—channel depth/channel width used only in laminar flow (dimensionless)
$x_{(FC)}$, $\Delta_{(FC)}$, $B_{(FC)}$—variables
$y$—switching function
$e$—exponential function
$ln$—natural logarithm
$FC$—index related to Fuel Cells

For $2^{16} = 65536$ Sobol Quasi Monte-Carlo pairs [48], which cover $Re_{FC} = 50$–4100 and for $h/H$ from 0.83 to 2.45, the maximal relative error of the final calculated flow friction factor $\lambda_{FC}$ using Equation (12) is 0.46% compared with the original Equation (2). The accuracy and speed of execution are tested through the code given in the next section.

## 3. Software Code and Measurement of Execution Speed

The unified equation for laminar and turbulent fan-induced air flow through open-cathode side of the observed PEMFCs is given in Equation (12), where the algorithm from Figure 2 should be followed.

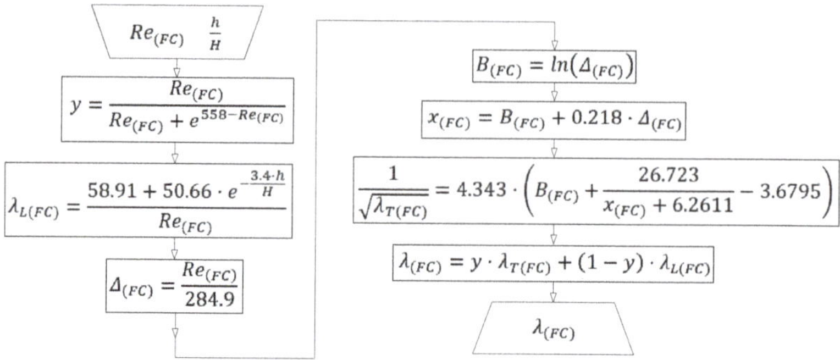


**Figure 2.** Algorithm for solving Equation (12).

The turbulent flow friction $\lambda_{T(FC)} \sim LT$ (with the intermediate step through $1/\sqrt{\lambda_{T(FC)}} \sim LT$) can be expressed using "wrightOmega" function as $LT = (B + wrightOmega(x) - x)/a$, but it can be executed faster using approximation as given in Equation (8), as $LT = 4.343 * (b + 26.723./(b + 0.218 * a + 6.2611) - 3.6795)$. The MATLAB code also works in GNU Octave, but it can be easily translated in any programming language. The final unified fuel cell model given by Equation (12) is coded in MATLAB as:

```
function L = PEMFCs(R,h)
a = log(10)/10;
d = R * a/65.6;
B = log(d);
x = B + 0.218 * d;
LT = 4.343 * (B + 26.723./(x + 6.2611) - 3.6795);
LT = 1./LT.^2;
LL = (58.91 + 50.66 * exp(-3.4 * h))./R;
y = R./(R + exp(558 - R));
L = y. * LT + (1 - y). * LL;
```

in this code:

Output parameters of the function:

-flow friction factor $\lambda_{(FC)}$ is given as $L$

Input parameters of the function:

-Reynolds number $Re_{(FC)}$ is given as $R$ (from the interval $50 < Re_{(FC)} < 4100$)

-channel depth/channel width $h/H$ is given as $h$ (from 0.83 to 2.5)

In MATLAB, symbol log() denotes the natural logarithm.

Implicitly given equations can be solved using iterative procedures [49,50], but also using appropriate accurate but also computationally efficient explicit approximations especially developed for the observed purpose. Our numerical results show that the computationally efficient approximation does not contain the time-consuming evaluation of the Wright $\omega$-function [6,7], but simple polynomials found by symbolic regression [51], which can be easily evaluated on computers. It is because simple functions such as $+$, $-$, $*$ and / are executed directly in the CPU and hence they are very fast [43].

Using $2^{16} = 65536$ Sobol Quasi Monte-Carlo pairs [48], which cover $Re_{(FC)} = 50$–4100 and for $h/H$ from 0.83 to 2.45, the evaluation of the MATLAB built-in "wrightOmega" function required 30.9 s, while our novel approximation required only 0.0022 s (our approximation is around fourteen thousand times faster). Consequently, numerical tests show that the novel approximation presented here is very suitable for modeling of fan-induced flow friction in a mesh with virtual roughness for air-forced flow in the open-cathode PEMFCs, as it is very fast and still very accurate.

## 4. Conclusions

This paper gives a novel numerically stable explicit solution for flow friction during airflow in cathode side of open-cathode PEMFCs. Symbolic regression is successfully used for obtaining a cheap but still accurate approximation of the Wright-$\omega$ function for fuel cells-based explicit friction model and also for approximations of the switching function, which is needed for the unified formulation of fuel cell friction model. Numerical experiments indicate that the novel approximation of the Wright-$\omega$ function reduced the computational time from half of a minute to fragments of a second. The relative error of the estimated friction flow factor is less than 0.5%.

In future research, further analyses and experiments are foreseen to show if and how this value of the virtual roughness $\varepsilon_{(FC)}$ of fan-induced friction in a mesh of conduits for air-forced flow in the open-cathode PEMFCs can be changed and how it depends on fuel cells parameters (type of fuel cells, size, geometry of channels, etc.). The study should be extended to cover other types of fuel cell, to include water and heat management, etc. [52–60].

**Author Contributions:** D.B. had the initial idea for this article and he dealt with physical interpretation of the problem. P.P. conducted detailed numerical experiments (included symbolic regression) following initial inputs of D.B. D.B. wrote a draft of this article. Both authors contributed equally to this article. All authors have read and agreed to the published version of the manuscript.

**Funding:** The work of D.B. has been supported by the Ministry of Education, Youth and Sports of the Czech Republic through the National Programme of Sustainability (NPS II) project "IT4Innovations excellence in science-LQ1602", while the work of P.P. has been partially supported by the Technology Agency of the Czech Republic under the project "Energy System for Grids" TK02030039. Additionally, the work of D.B. has been partially supported by the Ministry of Education, Science, and Technological Development of the Republic of Serbia.

**Conflicts of Interest:** The authors declare no conflict of interest.

## Notations

The following symbols are used in this article:

| For pipes: | |
|---|---|
| $\lambda_{T(p)}$ | turbulent Darcy flow friction factor for pipes (dimensionless) |
| $\lambda_{L(p)}$ | turbulent Darcy flow friction factor for pipes (dimensionless) |
| $Re_{(p)}$ | Reynolds number (dimensionless)—the same definition as for fuel cells |
| $\varepsilon_{(p)}$ | relative roughness of inner pipe surface (dimensionless) |
| $p$ | index related to pipes |
| For Fuel Cells: | |
| $\lambda_{(FC)}$ | unified Darcy flow friction factor for fuel cells (dimensionless) |
| $\lambda_{T(FC)}$ | turbulent Darcy flow friction factor for fuel cells (dimensionless) |
| $\lambda_{L(FC)}$ | laminar Darcy flow friction factor for fuel cells (dimensionless) |
| $Re_{(FC)}$ | Reynolds number (dimensionless)—the same definition as for pipes |
| $\varepsilon_{(FC)}$ | virtual relative roughness of fuel cell (dimensionless) |
| $\frac{h}{H}$ | channel depth/channel width used only in laminar flow (dimensionless) |
| $x_{(FC)}, \Delta_{(FC)}, B_{(FC)}$ | variables |
| $a_{(FC)}, b_{(FC)}, c_{(FC)}$ | constants |
| $FC$ | index related to Fuel Cells |
| Functions: | |
| $log_{10}$ | logarithmic function with base 10 |
| $ln$ | natural logarithm |
| $e$ | exponential function |
| $W$ | Lambert function |
| $\omega$ | Wright function |
| $y$ | switching function |

## References


1. Díaz-Curiel, J.; Miguel, M.J.; Caparrini, N.; Biosca, B.; Arévalo-Lomas, L. Improving basic relationships of pipe hydraulics. *Flow Meas. Instrum.* **2020**, *72*, 101698. [CrossRef]
2. Colebrook, C.F. Turbulent flow in pipes with particular reference to the transition region between the smooth and rough pipe laws. *J. Inst. Civ. Eng.* **1939**, *11*, 133–156. [CrossRef]
3. Colebrook, C.F.; White, C. Experiments with fluid friction in roughened pipes. *Proc. R. Soc. Lond. Ser. A Math. Phys. Sci.* **1937**, *161*, 367–381. [CrossRef]
4. Moody, L.F. Friction factors for pipe flow. *Trans. ASME* **1944**, *66*, 671–684.
5. Hayes, B. Why W? On the Lambert W function, a candidate for a new elementary function in mathematics. *Am. Sci.* **2005**, *93*, 104–108. [CrossRef]
6. Brkić, D.; Praks, P. Accurate and efficient explicit approximations of the Colebrook flow friction equation based on the Wright $\omega$-function. *Mathematics* **2019**, *7*, 34. [CrossRef]
7. Praks, P.; Brkić, D. Review of new flow friction equations: Constructing Colebrook's explicit correlations accurately. *Rev. Int. Métodos Numér. Cálc. Diseño Ing.* **2020**, *36*. in press.
8. Barreras, F.; López, A.M.; Lozano, A.; Barranco, J.E. Experimental study of the pressure drop in the cathode side of air-forced open-cathode proton exchange membrane fuel cells. *Int. J. Hydrogen Energy* **2011**, *36*, 7612–7620. [CrossRef]
9. Brkić, D. Comments on "Experimental study of the pressure drop in the cathode side of air-forced open-cathode proton exchange membrane fuel cells" by Barreras et al. *Int. J. Hydrogen Energy* **2012**, *37*, 10963–10964. [CrossRef]
10. Smith, R.; Miller, J.; Ferguson, J. *Flow of Natural Gas through Experimental Pipelines and Transmission Lines*; US Bureau of Mines American Gas Association AGA: New York, NY, USA, 1956; p. 89.
11. Plascencia, G.; Díaz-Damacillo, L.; Robles-Agudo, M. On the estimation of the friction factor: A review of recent approaches. *SN Appl. Sci.* **2020**, *2*, 163. [CrossRef]
12. Kheirabadi, A.C.; Groulx, D. Cooling of server electronics: A design review of existing technology. *Appl. Therm. Eng.* **2016**, *105*, 622–638. [CrossRef]
13. Khalaj, A.H.; Halgamuge, S.K. A Review on efficient thermal management of air-and liquid-cooled data centers: From chip to the cooling system. *Appl. Energy* **2017**, *205*, 1165–1188. [CrossRef]
14. Soupremanien, U.; Le Person, S.; Favre-Marinet, M.; Bultel, Y. Tools for designing the cooling system of a proton exchange membrane fuel cell. *Appl. Therm. Eng.* **2012**, *40*, 161–173. [CrossRef]
15. Guo, H.; Wang, M.H.; Ye, F.; Ma, C.F. Experimental study of temperature distribution on anodic surface of MEA inside a PEMFC with parallel channels flow bed. *Int. J. Hydrogen Energy* **2012**, *37*, 13155–13160. [CrossRef]
16. Henriques, T.; César, B.; Branco, P.C. Increasing the efficiency of a portable PEM fuel cell by altering the cathode channel geometry: A numerical and experimental study. *Appl. Energy* **2010**, *87*, 1400–1409. [CrossRef]
17. Zhao, C.; Xing, S.; Chen, M.; Liu, W.; Wang, H. Optimal design of cathode flow channel for air-cooled PEMFC with open cathode. *Int. J. Hydrogen Energy* 2020. [CrossRef]
18. Li, C.; Liu, Y.; Xu, B.; Ma, Z. Finite Time thermodynamic optimization of an irreversible Proton Exchange Membrane Fuel Cell for vehicle use. *Processes* **2019**, *7*, 419. [CrossRef]
19. De Lorenzo, G.; Andaloro, L.; Sergi, F.; Napoli, G.; Ferraro, M.; Antonucci, V. Numerical simulation model for the preliminary design of hybrid electric city bus power train with polymer electrolyte fuel cell. *Int. J. Hydrogen Energy* **2014**, *39*, 12934–12947. [CrossRef]
20. Darjat; Sulistyo; Triwiyatno, A.; Sudjadi; Kurniahadi, A. Designing hydrogen and oxygen flow rate control on a solid oxide fuel cell simulator using the fuzzy logic control method. *Processes* **2020**, *8*, 154. [CrossRef]
21. Taner, T. A Flow Channel with Nafion Membrane Material Design of Pem Fuel Cell. *J. Therm. Eng.* **2019**, *5*, 456–468. [CrossRef]
22. Majlan, E.H.; Rohendi, D.; Daud, W.R.W.; Husaini, T.; Haque, M.A. Electrode for proton exchange membrane fuel cells: A review. *Renew. Sustain. Energy Rev.* **2018**, *89*, 117–134. [CrossRef]
23. Fly, A.; Thring, R.H. A comparison of evaporative and liquid cooling methods for fuel cell vehicles. *Int. J. Hydrogen Energy* **2016**, *41*, 14217–14229. [CrossRef]

24. Rahgoshay, S.M.; Ranjbar, A.A.; Ramiar, A.; Alizadeh, E. Thermal investigation of a PEM fuel cell with cooling flow field. *Energy* **2017**, *134*, 61–73. [CrossRef]
25. Topal, H.; Taner, T.; Naqvi, S.A.; Altınsoy, Y.; Amirabedin, E.; Ozkaymak, M. Exergy Analysis of a circulating fluidized bed power plant co-firing with olive pits: A case study of power plant in Turkey. *Energy* **2017**, *140*, 40–46. [CrossRef]
26. Brkić, D. Can pipes be actually really that smooth? *Int. J. Refrig.* **2012**, *35*, 209–215. [CrossRef]
27. Praks, P.; Brkić, D. Choosing the optimal multi-point iterative method for the Colebrook flow friction equation. *Processes* **2018**, *6*, 130. [CrossRef]
28. Praks, P.; Brkić, D. Advanced iterative procedures for solving the implicit Colebrook equation for fluid flow friction. *Adv. Civ. Eng.* **2018**, *2018*, 5451034. [CrossRef]
29. Brkić, D. Review of explicit approximations to the Colebrook relation for flow friction. *J. Pet. Sci. Eng.* **2011**, *77*, 34–48. [CrossRef]
30. Brkić, D.; Praks, P. Unified friction formulation from laminar to fully rough turbulent flow. *Appl. Sci.* **2018**, *8*, 2036. [CrossRef]
31. Barreras, F.; Lozano, A.; Barranco, J.E. Response to the comments on "Experimental study of the pressure drop in the cathode side of air-forced open-cathode proton exchange membrane fuel cells" by Dejan Brkić. *Int. J. Hydrogen Energy* **2012**, *37*, 10965. [CrossRef]
32. Sharp, W.W.; Walski, T.M. Predicting internal roughness in water mains. *J. AWWA* **1988**, *80*, 34–40. [CrossRef]
33. Guo, X.; Wang, T.; Yang, K.; Fu, H.; Guo, Y.; Li, J. Estimation of equivalent sand–grain roughness for coated water supply pipes. *J. Pipeline Syst. Eng. Pract.* **2020**, *11*, 04019054. [CrossRef]
34. Bhui, A.S.; Singh, G.; Sidhu, S.S.; Bains, P.S. Experimental investigation of optimal ED machining parameters for Ti-6Al-4V biomaterial. *Facta Univ. Ser. Mech. Eng.* **2018**, *16*, 337–345. [CrossRef]
35. Niazkar, M.; Talebbeydokhti, N.; Afzali, S.H. Novel grain and form roughness estimator scheme incorporating artificial intelligence models. *Water Resour. Manag.* **2019**, *33*, 757–773. [CrossRef]
36. Niazkar, M.; Talebbeydokhti, N.; Afzali, S.H. Development of a new flow-dependent scheme for calculating grain and form roughness coefficients. *KSCE J. Civ. Eng.* **2019**, *23*, 2108–2116. [CrossRef]
37. Andersson, J.; Oliveira, D.R.; Yeginbayeva, I.; Leer-Andersen, M.; Bensow, R.E. Review and comparison of methods to model ship hull roughness. *Appl. Ocean Res.* **2020**, *99*, 102119. [CrossRef]
38. Keady, G. Colebrook-White formula for pipe flows. *J. Hydraul. Eng.* **1998**, *124*, 96–97. [CrossRef]
39. Brkić, D. Lambert W function in hydraulic problems. *Math. Balk.* **2012**, *26*, 285–292. Available online: http://www.math.bas.bg/infres/MathBalk/MB-26/MB-26-285-292.pdf (accessed on 21 May 2020).
40. Sonnad, J.R.; Goudar, C.T. Constraints for using Lambert W function-based explicit Colebrook–White equation. *J. Hydraul. Eng.* **2004**, *130*, 929–931. [CrossRef]
41. Brkić, D. Comparison of the Lambert W-function based solutions to the Colebrook equation. *Eng. Comput.* **2012**, *29*, 617–630. [CrossRef]
42. Sonnad, J.R.; Goudar, C.T. Turbulent flow friction factor calculation using a mathematically exact alternative to the Colebrook–White equation. *J. Hydraul. Eng.* **2006**, *132*, 863–867. [CrossRef]
43. Biberg, D. Fast and accurate approximations for the Colebrook equation. *J. Fluids Eng. Trans. ASME* **2017**, *139*, 031401. [CrossRef]
44. Muzzo, L.E.; Pinho, D.; Lima, L.E.; Ribeiro, L.F. Accuracy/speed analysis of pipe friction factor correlations. In Proceedings of the International Congress on Engineering and Sustainability in the XXI Century 2019, Faro, Portugal, 9–11 October 2019; pp. 664–679. [CrossRef]
45. Zeyu, Z.; Junrui, C.; Zhanbin, L.; Zengguang, X.; Peng, L. Approximations of the Darcy–Weisbach friction factor in a vertical pipe with full flow regime. *Water Supply* **2020**. [CrossRef]
46. Dubčáková, R. Eureqa: Software review. *Genet. Program. Evol. M.* **2011**, *12*, 173–178. [CrossRef]
47. Schmidt, M.; Lipson, H. Distilling free-form natural laws from experimental data. *Science* **2009**, *324*, 81–85. [CrossRef]
48. Wagner, S.; Kronberger, G.; Beham, A.; Kommenda, M.; Scheibenpflug, A.; Pitzer, E.; Vonolfen, S.; Kofler, M.; Winkler, S.; Dorfer, V.; et al. Architecture and design of the HeuristicLab optimization environment. *Top. Intell. Eng. Inform.* **2014**, *6*, 197–261. [CrossRef]

49. Sobol, I.M.; Turchaninov, V.I.; Levitan, Y.L.; Shukhman, B.V. *Quasi-Random Sequence Generators*; Distributed by OECD/NEA Data Bank; Keldysh Institute of Applied Mathematics; Russian Academy of Sciences: Moscow, Russia, 1992. Available online: https://ec.europa.eu/jrc/sites/jrcsh/files/LPTAU51.rar (accessed on 20 May 2020).
50. Winning, H.K.; Coole, T. Explicit friction factor accuracy and computational efficiency for turbulent flow in pipes. *Flow Turbul. Combust.* **2013**, *90*, 1–27. [CrossRef]
51. Winning, H.K.; Coole, T. Improved method of determining friction factor in pipes. *Int. J. Numer. Methods Heat Fluid Flow* **2015**, *25*, 941–949. [CrossRef]
52. Praks, P.; Brkić, D. Rational Approximation for Solving an Implicitly Given Colebrook Flow Friction Equation. *Mathematics* **2020**, *8*, 26. [CrossRef]
53. Taner, T. Energy and Exergy Analyze of PEM fuel cell: A case study of modeling and simulations. *Energy* **2018**, *143*, 284–294. [CrossRef]
54. Andaloro, L.; Napoli, G.; Sergi, F.; Dispenza, G.; Antonucci, V. Design of a hybrid electric fuel cell power train for an urban bus. *Int. J. Hydrogen Energy* **2013**, *38*, 7725–7732. [CrossRef]
55. Taner, T. Alternative energy of the future: A technical note of PEM fuel cell water management. *J. Fundam. Renew. Energy Appl.* **2015**, *5*, 1000163.
56. Andaloro, L.; Arista, A.; Agnello, G.; Napoli, G.; Sergi, F.; Antonucci, V. Study and design of a hybrid electric vehicle (Lithium Batteries-PEM FC). *Int. J. Hydrogen Energy* **2017**, *42*, 3166–3184. [CrossRef]
57. Taner, T. The micro-scale modeling by experimental study in PEM fuel cell. *J. Therm. Eng.* **2017**, *3*, 1515–1526. [CrossRef]
58. Napoli, G.; Micari, S.; Dispenza, G.; Di Novo, S.; Antonucci, V.; Andaloro, L. Development of a fuel cell hybrid electric powertrain: A real case study on a Minibus application. *Int. J. Hydrogen Energy* **2017**, *42*, 28034–28047. [CrossRef]
59. Taner, T.; Naqvi, S.A.; Ozkaymak, M. Techno-Economic Analysis of a more efficient hydrogen generation system prototype: A case study of PEM electrolyzer with Cr-C coated Ss304 bipolar plates. *Fuel Cells* **2019**, *19*, 19–26. [CrossRef]
60. Mendicino, G.; Colosimo, F. Reply to Comment by J. Qin and T. Wu on "Analysis of flow resistance equations in gravel bed rivers with intermittent regimes: Calabrian fiumare data set". *Water Resour. Res.* **2020**, *56*, e2019WR027003. [CrossRef]

*Article*

# Environmental Sustainability Assessment of Typical Cathode Materials of Lithium-Ion Battery Based on Three LCA Approaches

**Lei Wang [1], Haohui Wu [1], Yuchen Hu [1], Yajuan Yu [1,*] and Kai Huang [2,*]**

1 School of Materials Science and Engineering, Beijing Institute of Technology, Beijing 100081, China; 3120181158@bit.edu.cn (L.W.); 2120171208@bit.edu.cn (H.W.); 3220180937@bit.edu.cn (Y.H.)
2 College of Environmental Science and Engineering, Beijing Forestry University, Beijing 100083, China
* Correspondence: 04575@bit.edu.cn (Y.Y.); huangkai@bjfu.edu.cn (K.H.)

Received: 2 January 2019; Accepted: 24 January 2019; Published: 7 February 2019

**Abstract:** With the rapid increase in production of lithium-ion batteries (LIBs) and environmental issues arising around the world, cathode materials, as the key component of all LIBs, especially need to be environmentally sustainable. However, a variety of life cycle assessment (LCA) methods increase the difficulty of environmental sustainability assessment. Three authoritative LCAs, IMPACT 2002+, Eco-indicator 99(EI-99), and ReCiPe, are used to assess three traditional marketization cathode materials, compared with a new cathode model, $FeF_3(H_2O)_3/C$. They all show that four cathode models are ranked by a descending sequence of environmental sustainable potential: $FeF_3(H_2O)_3/C$, $LiFe_{0.98}Mn_{0.02}PO_4/C$, $LiFePO_4/C$, and $LiCoO_2/C$ in total values. Human health is a common issue regarding these four cathode materials. Lithium is the main contributor to the environmental impact of the latter three cathode materials. At the midpoint level in different LCAs, the toxicity and land issues for $LiCoO_2/C$, the non-renewable resource consumption for $LiFePO_4/C$, the metal resource consumption for $LiFe_{0.98}Mn_{0.02}PO_4/C$, and the mineral refinement for $FeF_3(H_2O)_3/C$ show relatively low environmental sustainability. Three LCAs have little influence on total endpoint and element contribution values. However, at the midpoint level, the indicator with the lowest environmental sustainability for the same cathode materials is different in different methodologies.

**Keywords:** LIBs; environmental sustainability; cathode material; LCA

## 1. Introduction

With the expansion of the LIBs market, new cathode materials are constantly being developed [1]. In terms of weight fraction and cost, the cathode part for LIB is the most significant sector [2]. However, long-standing effort has been devoted to the development of high energy density and capability cathode materials [3], meeting the demand of electric vehicles, power tools, and large electric power storage units [4]. In fact, with the increase in energy density and capacity, many trace LIBs have an increasing impact on the environment [5]. Meanwhile, modern society must overcome many difficulties, such as obtaining natural resources and protecting the natural environment [4]. Before we commercialize a new cathode model, its environmental cost should be considered. Andersson confirmed the feasibility of environmental sustainability assessment in LCA for product development [6]. Numerous studies have quantified the impact of different types of LIB on the production process in its lifecycle [7]. Slowly, LCA is becoming more commonly used as a standardized method to determine the impact of a product or service on the environment throughout its whole life [8].

Although many new cathode models are being researched, the introduction of $LiCoO_2/C$ has enabled the commercialization of the first LIB [2]. LIBs have been available on the market from Sony

Corp. since the early 1990s, and $LiCoO_2/C$ has become the leading LIB system [9]. $LiFePO_4/C$ stands as a competitive candidate cathode material for the next generation of a green and sustainable LIB system due to its long life span, abundant resources, low toxicity, and high thermal stability [10]. Meanwhile, the $LiFe_{0.98}Mn_{0.02}PO_4/C$, as an improved cathode material for $LiFePO_4/C$ [11], also shows promising development potential. As Zeng [12] confirmed, the electrochemical performance of $LiFePO_4/C$ was remarkably improved by a slight manganese substitution, creating the general formula $LiFe_XMn_{1\text{-}X}PO_4/C$ [13]. However, most commercial cathode materials are $LiCoO_2/C$, whose actual capacity is 140–155 $mAhg^{-1}$, while $LiFePO_4/C$ or $LiFe_{0.98}Mn_{0.02}PO_4/C$'s theoretical capacity is only 170 $mAhg^{-1}$, so LIBs urgently need a new cathode model, like $FeF_3(H_2O)_3/C$ [14].

We can see that electrochemical properties [15,16] have been the driving force for the development of different cathode materials. Many new studies have focused on electrochemical properties. However, due to the increasing prevalence of different cathode materials in electronic equipment and vehicles, their impact on the environment also needs to be considered [17]. Peters summarized the reviewed studies, focusing on energy demand and warming gas emissions for $LiCoO_2/C$ and $LiFePO_4/C$ [7]. Wang [18] evaluated the LIB with lithium-rich materials used in an electric vehicle throughout the life cycle of the battery. Wang [19] measured the carbon footprints of three industry lithium-ion secondary battery chains, and came to the conclusion that electric energy consumption is the main factor of lithium ion battery production companies in generating a carbon footprint. Liang [20] directly adopted LCA to assess the greenhouse gas emissions of LIBs. Deng [21] used LCA to assess high capacity molybdenum disulfide LIB for electric vehicles. Zackrission [22] elevated lithium-air batteries by LCA to quantify its climate impact, abiotic resource depletion and toxicity. Gong [23] evaluated four cathode materials by family footprint including carbon footprint and water footprint, and both direct and indirect water footprint [24].

In this paper, we not only pay attention to the total environmental sustainability of these four cathode materials, as previous studies have done, but also find that different indicators damage their environmental sustainability in different ways. Indeed, except for these common problems, certain indicators play different roles in specific cathode materials. In addition, from the results, we find that the different LCAs we choose have impacts on our evaluation. Based on the analysis of the calculation in three kinds of LCA, some suggestions are put forward for different cathodes, and the influence of different LCA is studied. There are three LCAs we chose for our study. Firstly, IMPACT 2002+ life cycle impact assessment methodology proposes a feasible implementation of a combined midpoint/damage approach, linking all types of life cycle inventory results (elementary flows and other interventions) via 14 midpoint categories to four damage categories, especially to human toxicity and ecotoxicity [25]. Secondly Eco-indicator 99(EI-99) is also used, a damage-oriented method to assess the emissions, extractions, and land use in all processes, and the damage to human health, ecosystem quality, and resources is calculated [26] ReCiPe assesses 18 impact categories at midpoint level (ozone depletion, human toxicity, etc.), and three endpoint categories (human health, ecosystems, resources) at endpoint level [27]. Based on these three LCAs, we divide the results into four innovative parts: (1) according to the total value of four different cathode materials, we have a preliminary understanding of the environmental sustainability ranking of these four cathode materials; (2) according to the endpoint value, we find one common environmental problem of the four cathode materials; (3) from the midpoint value, indicators showing the best or worst environmental sustainability of four cathode materials are summarized; (4) from the perspective of element contribution, the key elements that have obvious influence on the environmental sustainability of these four cathode materials are found. In fact, $LiCoO_2/C$ does not always show the lowest environmental sustainability among all indicators. The new cathode model $FeF_3(H_2O)_3/C$ also shows the lowest environmental sustainability in some respects. Although the chemical composition of $LiFe_{0.98}Mn_{0.02}PO_4/C$ is similar to that of $LiFePO_4/C$, the metal resource consumption for the former is far larger. Instead of focusing on the overall environmental sustainability, we care more about these specific indicators in different LCAs.

## 2. Materials and Methods

### 2.1. LCAs and Cathode Materials

At present, IMPACT 2002+ and EI-99 have been widely recognized in many studies used to quantify the environmental impact of products [28]. Moreover, ReCiPe, as an improved LCA for EI-99, can evaluate more detailed indicators [29]. These three LCA methods are selected as the research methods. Although the results of environmental impacts are similar for equivalent categories, the ReCiPe and IMPACT 2002+ methods provide more categories for evaluation and comparison than EI-99 [30]. Furthermore, the assessment categories among these three LCAs are sufficient for us to make a comprehensive comparison. For example, the water issue is only evaluated in IMPACT 2002+, and acidification issues are only considered in both IMPACT 2002+ and EI-99. Indeed, there are many differences among these three LCAs, like the characterization of the unit and categories [31]. In addition, three cathode materials of LIB industry, $LiCoO_2/C$, $LiFePO_4/C$, and $LiFe_{0.98}Mn_{0.02}PO_4/C$, are selected as research objects. These three cathode models are marketed and mass produced every year. The evaluation of these major cathode materials is of great significance to the environmental sustainability development of LIBs. Furthermore, a new cathode model, $FeF_3(H_2O)_3/C$, is also evaluated as a comparison. The discovery of its environmental advantages can provide some enlightenment for these three market-oriented cathode models, which will help us to manufacture environmentally friendly cathode materials. Finally, the impact of different LCAs on sustainability assessment is also considered.

### 2.2. Experimental Designation

#### 2.2.1. Scope and Function Unit

To compare the four cathode materials on the same basis, we should stipulate the scope and function unit of these four cathode materials. We only research the environmental impact of the cathode part. The original quality list comes from the existing literature and laboratory. After a normalized conversion, we make the quality list meet the functional unit (1 kg), which means that the mass of each cathode model we evaluated is equal.

#### 2.2.2. Experimental Devices

This study is conducted by global LCA software Simapro released in 1990, and more than 80 countries have recognized its authority. Simapro allows researchers to collect, analyzes, and monitor the sustainability performance of products and services, from extraction of raw materials to manufacturing, distribution, use, and disposal [32]. In general, this includes (a) a user interface for modeling the product system, (b) a life cycle unit process database, (c) an impact assessment database with data supporting several life cycle impact assessment methodologies, and (d) a calculator that combines numbers from the databases in accordance with the modeling of the product system in the user interface [33].

#### 2.2.3. Experimental Process

We researched these four cathodes at a deeper level. Of course, the total value of environmental sustainability for these four cathodes is shown in three LCAs. However, we also looked at endpoint values and midpoint values. In short, the lower the value is, the better the environmental sustainability. Finally, the element contribution proportion to endpoint and total values is calculated.

(1) Simulated Assembly

To meet the requirements of functional unit (1 kg), the original data were converted into a standard quality list. Then we assembled four complete cathode models in Simapro.

(2) Calculation

To compare these four cathode materials, we have processed on the raw calculated data included in Supplementary Materials (Tables S1–S3), like normalization and contribution analysis. In fact, the data in the Supplementary Materials is calculated firstly by Simapro and all values has their own units which undoubtedly make it more difficult to compare these cathode materials. The detailed calculations of these four cathodes are listed as follows:

i. we calculate the total value of the four cathode materials by the entropy method, so that it satisfies a unit (Pt).
ii. we calculate the values of all endpoint indicators in the unit (Pt) and calculate their contribution to the total value.
iii. we calculate the values of all midpoint indicators and give the normalized values of all midpoint indicators in this paper. Material B has been a major market cathode in recent years, especially in the field of electric vehicles. We choose all values of midpoint indicators of material B as baseline 1.
iv. we calculate the contribution ratio of elements at the endpoint and the total level.

Some indicators are simplified: the total value in three LCAs, Ti(i = 1, 2, 3); the endpoint indicators in different LCAs(IMPACT 2002+, Xm(m = 1, 2, 3, 4); EI-99, Yn(n = 1, 2, 3); ReCiPe, Zh(h = 1, 2, 3)). The research process is shown in Figure 1.

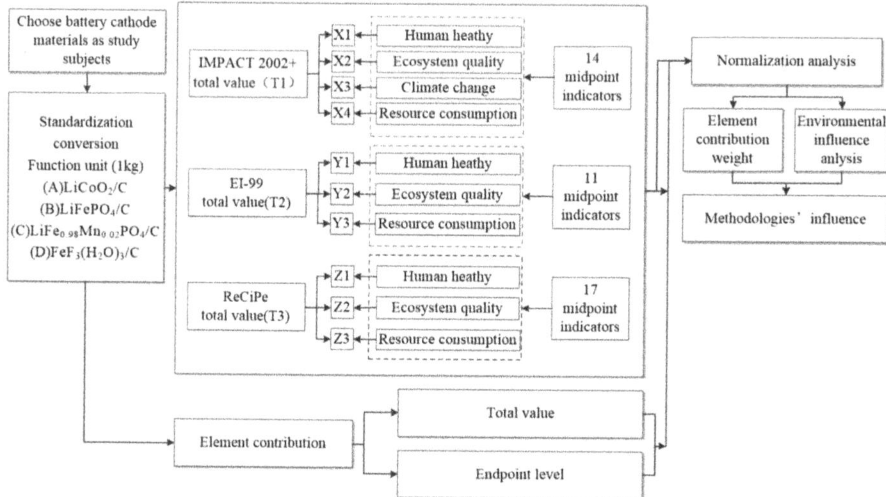


**Figure 1.** The research process.

## 3. Results and Discussion

### 3.1. The Total Value of the Four Cathodes' Environmental Impact

It is necessary to qualify the total environmental impact of these four cathode materials before a deeper investigation, because all midpoint and endpoint indicators are divided from the total environmental impact. Unlike Gong's research on evaluating environmental, economic, and electrochemical performance indicators by footprint family and Peters's summary focusing on energy demand and warming gas emissions for $LiCoO_2/C$ and $LiFePO_4/C$, we make a pure and further environmental assessment for these four cathode materials, and do not consider the economic benefits, electrochemical properties, energy demand, etc. This study aims to make a comprehensive assessment of these four typical cathodes directly. The total values for these four cathode materials in three LCAs are shown in Figure 2.

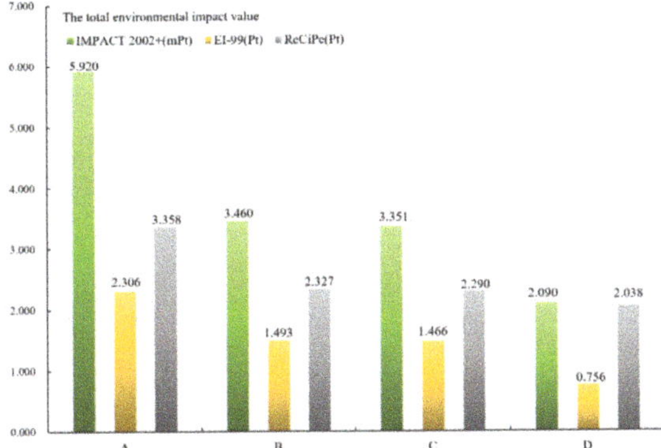


**Figure 2.** Total values for four cathodes in three LCAs: (**A**) $LiCoO_2/C$, (**B**) $LiFePO_4/C$, (**C**) $LiFe_{0.98}Mn_{0.02}PO_4/C$, (**D**) $FeF_3(H_2O)_3/C$.

As Figure 2 shows, we sort the four cathode types in descending order of environmental sustainability potential: D, C, B, and A. Although different LCAs have their own calculation standards, the trend of these four cathodes is the same. As Gong [23] confirms, D has the best environmental performance compared with B and C. However, the study does not explain the differences in specific indicators between B, C, and D. More detailed indicators that are important for each of the three cathodes are found in this study. In addition, material A is added to the study as another major market cathode model. No matter what LCA we choose, the new cathode model D always presents the best potential for environmental sustainability, while material A performs the worst. The environmental sustainability of material B is close to that of material C. Methodological emphasis on environmental assessment is only reflected in quantitative values, rather than the qualitative environmental sustainable potential among these four cathodes.

### 3.2. Endpoint Level

To distinguish concrete environmental impacts of different cathode materials, we calculate all endpoint indicators in Table 1. In any LCA, material A always has the highest endpoint value among these four materials. D is the smallest. Except the resource consumption in ReCiPe, 0.449 Pt of C larger than 0.393 Pt of B, other values for material B are slight larger than those for material C. IMPACT 2002+ and EI-99 have less impact on these four cathode materials' ranking of environmental sustainable potential. The resource consumption value between B and C calculated by ReCiPe is different to that calculated by the other two LCAs.

**Table 1.** Endpoint values for four cathodes.

| Endpoint | Indicator Value/Pt | | | |
|---|---|---|---|---|
| | A | B | C | D |
| X1 | $2.510\times10^{-03}$ | $1.443\times10^{-03}$ | $1.423\times10{-}03$ | $0.779\times10^{-03}$ |
| X2 | $0.966\times10^{-03}$ | $0.093\times10^{-03}$ | $0.090\times10{-}03$ | $0.068\times10^{-03}$ |
| X3 | $1.278\times10^{-03}$ | $0.937\times10^{-03}$ | $0.900\times10{-}03$ | $0.555\times10^{-03}$ |
| X4 | $1.166\times10^{-03}$ | $0.988\times10^{-03}$ | $0.938\times10{-}03$ | $0.688\times10^{-03}$ |
| Y1 | 1.796 | 1.142 | 1.128 | 0.528 |
| Y2 | 0.101 | 0.042 | 0.041 | 0.035 |
| Y3 | 0.409 | 0.309 | 0.297 | 0.193 |
| Z1 | 2.697 | 1.875 | 1.784 | 1.747 |
| Z2 | 0.134 | 0.060 | 0.058 | 0.034 |
| Z3 | 0.527 | 0.393 | 0.449 | 0.258 |

Note: refer all endpoints (X1-4; Y1-2; Z1-3) to Figure 1.

The total value consists of the values of three or four endpoint indicators. To make a more intuitive observation, we calculated the contribution rates of different endpoint indexes to the total value, as shown in Figure 3. In three LCAs, the maximum contribution proportion of these four cathodes comes from human health. For material A, the contribution rate of ecosystem quality is relatively large in IMPACT 2002+.

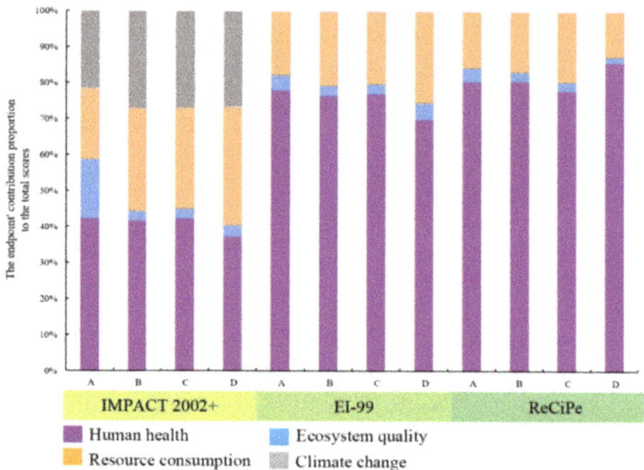


**Figure 3.** Endpoint indicators' contribution to the total value.

Obviously, in three LCAs, the environmental sustainability capacity for these four cathodes is related to human health, which means human health is a common problem for these four cathode materials to solve. In fact, Wanger [34] has confirmed that the effect of LIBs on human health is a common problem for LIBs. For the cathode, the effect on human health remains a major concern. Its existence may require a major technical improvement to overcome. For these individual problems in different cathodes, we can learn from the strengths and weaknesses of different cathodes. For example, material A always shows the largest environmental load in these four cathodes. Reducing its yield or finding alternative models, elements, or mechanisms is a feasible way to reduce its impact on ecosystem quality. As we know, IMPACT 2002+ and EI-99 have the same ranking for the environmental sustainability among these four cathode materials. Three LCAs show that the impact on human health is a common problem for these four cathode materials. However, in ReCiPe, material C consumes more resources compared with B. In IMPACT 2002+, material A's impact on ecosystem quality makes a relative contribution to its total environmental impact.

### 3.3. Midpoint Level

Similarly, each endpoint indicator can be divided into a number of midpoint indicators. In order to avoid interference from different units and magnitude, we choose all values of material B as the benchmark and normalize all values of the other three cathode materials. These indicators with an extreme value always show great disadvantages and advantages for different cathodes, and these normalized values, obviously larger or less than 1, are more meaningful for individual cathode improvement. In IMPACT 2002+, the difference values between all normalized values and B′ normalized value (1) are shown in Figure 4. In order to consider the impact of different LCA, we still give the unit of each midpoint indicator, reflecting their evaluation criteria. As we can see, there are 14 midpoint indicators, each of which has a different unit of measurement. To some extent, IMPACT 2002+ is more suitable for characterization evaluation. For example, we can use unit kg $PO_4$ p-lim to express the land use problem. Moreover, because of the presence of phosphorus, the data are meaningful for eutrophication.

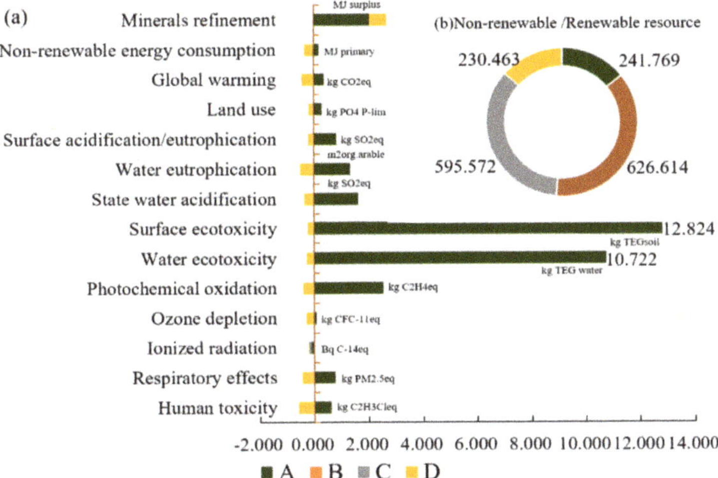


**Figure 4.** The normalized midpoint value in IMPACT 2002+.

As Figure 4 shows, when we regarded all normalized values of B as the baseline, except for the value of ionized radiation, other normalized values of A are larger than those of B. In particular, the eco-toxicity values of water body and surface are much larger, about 10.722 and 12.824, respectively. The high toxicity of cobalt may be the main reason for this situation. In fact, water problems and surface problems are difficult to completely separate, e.g., the toxicity of surface water. For water toxicity of A, water footprint assessment [35] may be a great method to quantify its water problems and reflect its toxicity from another perspective. For material D, except for the value for mineral refinement, other values are less than for B. Though D shows the best environmental sustainability, a green mineral refinement process is needed for material D. Finally, all normalized values for material C are slightly less than those for B. Material C, as an improved cathode to B by slight manganese substitution, has similar environmental sustainability potentiality to material B. The close element composition between material B and C, as the common formula shows, $LiFe_XMn_{1-X}PO_4/C$ (x = 0.98 in this study), accounts for the similar environmental sustainability. IMPACT 2002+ shows a sensitive assessment for cathode A, especially on water and surface ecotoxicity.

In EI-99, normalized values are shown in Figure 5. These indicators have the same cells separated from the same endpoint indicators. Compared with the IMPACT 2002+, these indicators cannot reflect specific substances due to their common units.

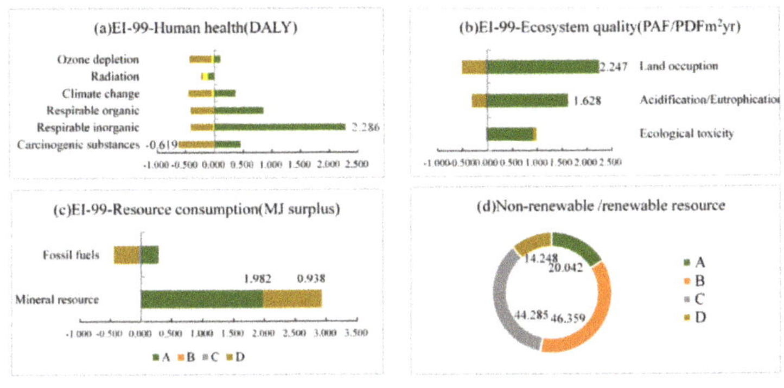


**Figure 5.** The normalized midpoint value in EI-99.

As Figure 5a–c shows, except for the radiation value, the normalized value of A is greater than that of B, and the respiratory inorganic matter, land occupation, and mineral resource problems of A are obvious, at 2.286, 2.247, and 1.982, respectively. For D, its mineral resource value is far larger, about 0.938. The ecological toxicity value is slightly larger. In addition to the mineral problems noted in IMPACT 2002+, the ecological toxicity of material D in EI-99 also becomes a low environmental sustainability index. Finally, all values of C are close to 1, as IMPACT 2002+ shows.

In ReCiPe, all normalized values are shown in Figure 6. The midpoint indicators in ReCiPe are more detailed than EI-99. The unit for these indictors is different.

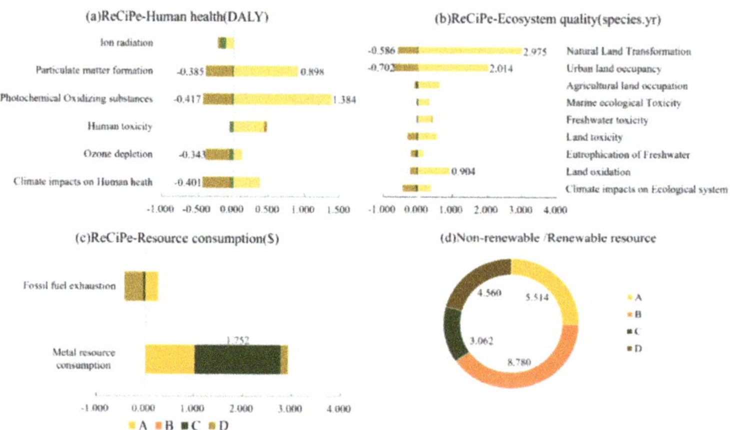


**Figure 6.** The normalized midpoint value in ReCiPe.

As Figure 6a–c shows, the normalized values of nature land transformation and urban land occupancy for material A are far larger, at 2.975 and, 2.014 respectively. Actually, material D performs very well on these two indicators, with −0.586 and −0.702, respectively. Land occupation is an important part of the assessment of ecosystem quality [36]. That is a key difference between material A and D. For material C, not all normalized values are close to B, especially metal resource consumption, which is as high as 1.752. The method ReCiPe concentrates more on the economic costs ($) in resource consumption [37], which means the economic cost in the slight manganese substitution process for material B needs to be cut down. To reduce the consumption of metal resources in the manganese substitution process must be a key issue for the development of material C. Finally, material D had low environmental sustainability in human toxicity, freshwater toxicity, and metal consumption compared with material B.

In addition, the problem of resource consumption deserves our attention. This endpoint in three LCAs is divided into two common midpoints, renewable and non-renewable resource consumption. We calculated the ratio of non-renewable resources to renewable ones. Mineral refinement in IMPACT 2002+, the mineral resource in EI-99 and the metal resource in ReCiPe are regarded as the renewable resource consumption. Likewise, the non-renewable energy, the fossil fuels and the fuel exhaustion are divided into the non-renewable resource consumption. The ratio is the non-renewable resources consumption per unit renewable resources consumption.

As Figures 4b, 5d and 6d show, material B always has the highest ratio in three LCAs, 626.614 (MJ primary/MJ surplus) in IMPACT 2002+, 41.359 (MJ/MJ) in EI-99, 8.780 ($/$) in ReCiPe. That means every unit renewable resource consumption needs more non-renewable resource in the whole life cycle of material B. Material B has the lowest environmental sustainability among these four cathodes. More green processes with low non-renewable resource consumption are needed for material B. As worldwide concern about fossil fuels grows, efforts at non-renewable resource protection are urgently required [38]. These high ratios need to be reduced. Integrating the renewable resources in a small isolated power system, an isolated and complete battery [39], and improving the capacity for cathodes [40] are promising directions to achieve this goal.

Among the four cathode materials, the emphasis in the three LCAs is different. For material A, three LCAs all think that iron radiation is not a serious issue. The main problem in IMPACT 2002+ is ecotoxicity. On the contrary, EI-99 and ReCiPe think that the land issue is a serious issue. For materials B and C, the values are mostly close to each other except for the metal resource consumption. For material D, three LCAs all show its low environmental sustainability in terms of mineral resource consumption, and its toxicity is noted in EI-99 and ReCiPe.

### 3.4. The Element Contribution to Environmental Sustainability

In this part, we use the elemental symbol to represent all elements in tables as follows: lithium (Li), cobalt (Co), manganese (Mn), iron (Fe), fluorine (F), nitrogen (N), phosphorus (P), and oxygen (O). The contribution of the element to the endpoint and the total value is calculated.

#### 3.4.1. The Element Contribution in Endpoint Level

In the three LCAs, the element contribution rate of material A is shown in Table 2. Cobalt is the largest contributor to ecosystem quality, followed by lithium. The other endpoint indexes were mainly affected by lithium, followed by cobalt. So cobalt impairs the environmental sustainability of material A in terms of ecosystem quality (endpoint) and land and toxicity issues (midpoint).

**Table 2.** The element contribution proportion for material A.

| LCAs | Endpoint | Li | O | Co |
|---|---|---|---|---|
| IMPACT 2002+ (%) | X1 | 53.60 | 0.11 | 46.30 |
| | X2 | 8.30 | 0.01 | 91.70 |
| | X3 | 64.70 | 0.26 | 35.10 |
| | X4 | 66.70 | 0.36 | 33.00 |
| EI-99 (%) | Y1 | 62.60 | 0.05 | 37.40 |
| | Y2 | 30.70 | 0.06 | 69.30 |
| | Y3 | 62.50 | 0.23 | 37.30 |
| ReCiPe (%) | Z1 | 58.30 | 0.07 | 41.60 |
| | Z2 | 40.50 | 0.09 | 59.40 |
| | Z3 | 57.90 | 0.20 | 41.90 |

For material B, the contribution rate of elements is shown in Table 3. The highest contribution proportion of each endpoint value always comes from lithium (the largest value is Y1 (83.70%), and the smallest value is Y2 (60.50%)), followed by phosphorus. Oxygen, iron, and nitrogen contribute little to each endpoint value.

**Table 3.** The element contribution proportion for material B.

| LCAs | Endpoint | Li | O | Fe | N | P |
|---|---|---|---|---|---|---|
| IMPACT 2002+ (%) | X1 | 78.20 | 0.91 | 1.14 | 2.28 | 17.5 |
| | X2 | 72.40 | 0.28 | 3.09 | 3.31 | 20.9 |
| | X3 | 74.10 | 1.63 | 0.16 | 4.41 | 19.7 |
| | X4 | 66.00 | 1.96 | 0.16 | 5.73 | 26.2 |
| EI-99 (%) | Y1 | 83.70 | 0.42 | 1.81 | 1.34 | 12.7 |
| | Y2 | 60.50 | 0.81 | 0.40 | 3.18 | 35.1 |
| | Y3 | 69.50 | 1.40 | 0.40 | 4.44 | 24.3 |
| ReCiPe (%) | Z1 | 70.40 | 0.49 | 0.34 | 5.64 | 23.2 |
| | Z2 | 76.20 | 0.97 | 0.17 | 3.84 | 18.9 |
| | Z3 | 65.20 | 1.25 | 6.09 | 4.27 | 23.2 |

Table 4 shows the element contribution rate of material C. The highest contribution rate of element to each endpoint value is still lithium, followed by phosphorus. The rest of the elements have less weight.

**Table 4.** The element contribution proportion for material C.

| LCAs | Endpoint | Li | Fe | O | N | P | Mn |
|---|---|---|---|---|---|---|---|
| IMPACT 2002+ (%) | X1 | 79.30 | 1.16 | 0.93 | 0.18 | 17.70 | 0.72 |
| | X2 | 74.30 | 3.17 | 0.28 | 0.27 | 21.50 | 0.47 |
| | X3 | 77.00 | 0.17 | 1.71 | 0.36 | 20.50 | 0.23 |
| | X4 | 69.50 | 0.17 | 2.08 | 0.48 | 27.60 | 0.25 |
| EI-99 (%) | Y1 | 83.60 | 1.72 | 0.40 | 0.14 | 13.80 | 0.34 |
| | Y2 | 63.50 | 0.34 | 0.66 | 0.31 | 33.90 | 1.29 |
| | Y3 | 72.20 | 0.42 | 1.47 | 0.37 | 25.20 | 0.29 |
| ReCiPe (%) | Z1 | 73.90 | 0.36 | 0.52 | 0.47 | 24.30 | 0.37 |
| | Z2 | 78.50 | 0.18 | 1.00 | 0.31 | 19.50 | 0.53 |
| | Z3 | 57.00 | 5.33 | 1.10 | 0.30 | 20.30 | 16.00 |

Finally, the element contribution proportion for material D is showed in Table 5. As a non-lithium composition model (no lithium in quality list), fluorine has the highest contribution rate among all mid-point indicators, at almost 90.00%.

**Table 5.** The element contribution proportion for material D.

| LCAs | Endpoint | Fe | O | F |
|---|---|---|---|---|
| IMPACT 2002+ (%) | X1 | 1.78 | 1.47 | 96.80 |
| | X2 | 3.55 | 0.33 | 96.10 |
| | X3 | 0.23 | 2.39 | 97.40 |
| | X4 | 0.19 | 2.44 | 97.40 |
| EI-99 (%) | Y1 | 3.08 | 0.73 | 96.20 |
| | Y2 | 0.33 | 0.66 | 99.00 |
| | Y3 | 0.54 | 1.95 | 97.50 |
| ReCiPe (%) | Z1 | 0.31 | 0.46 | 99.20 |
| | Z2 | 0.25 | 1.47 | 98.30 |
| | Z3 | 7.81 | 1.64 | 90.50 |

Some issues can be found by the element contribution. The ecosystem quality in material A is mainly affected by cobalt. To seek a substitute material or reduce the quality of cobalt in production process will be a method to improve its environmental sustainability. As one heavy metal element, cobalt has higher ecosystem toxicity and pollution capacity than other elements in the four cathode materials. This is why many efforts to recover A are concentrated not only on lithium but also on cobalt [41]. There are two examples of Co substitutes. Xiang [42] improved the electrochemical kinetics of lithium manganese phosphate via Co-substitution. Di Lecce [43] investigated a new Sn-C/$LiFe_{0.1}Co_{0.9}PO_4$ full lithium-ion cell with ionic liquid-based electrolyte. The two studies demonstrated the feasibility of producing Co-substitutes and for improving the environmental sustainability for A.

Except the ecosystem quality for material A mainly affected by cobalt, others indicators' environmental sustainability for materials A, B and C is mainly affected by lithium. However, the phosphorus in material B and C also has a great impact on their environmental sustainability. Reducing the consumption of lithium is an ongoing aim in the development for LIBs. More research on cobalt and phosphorus is also needed. All three LCAs show the same result—that is, different LCAs have no influence on the determination of the main elements contributing to the endpoint index values.

#### 3.4.2. The Element Contribution to the Total Value

The element proportions in the total level are shown in Figure 7.

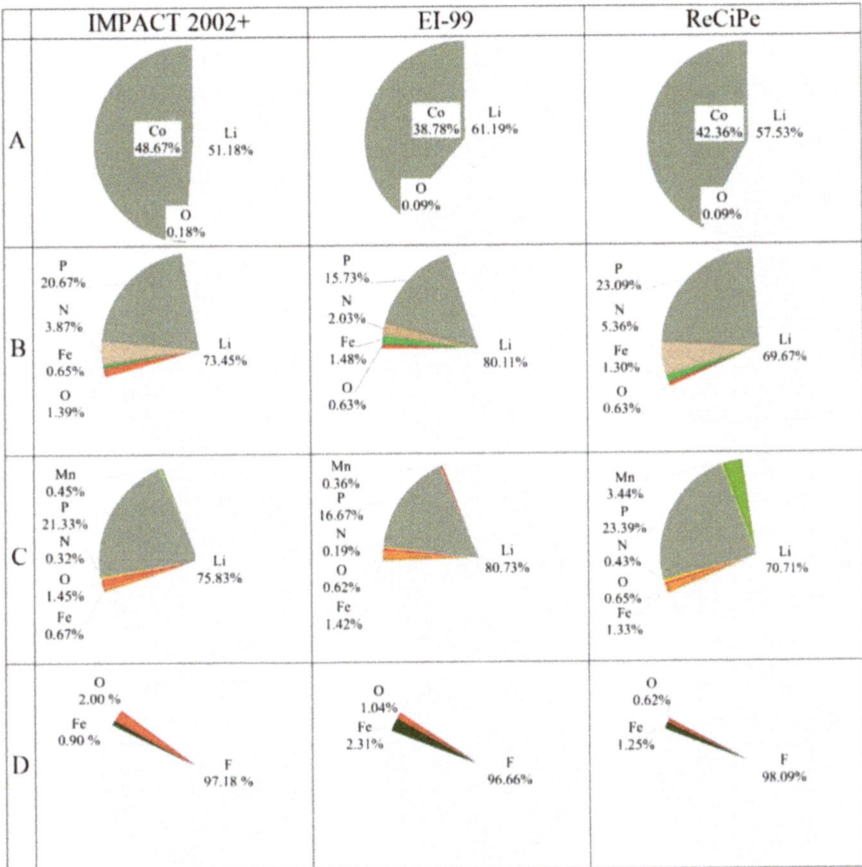


**Figure 7.** The element contribution proportion to the total values in three LCAs.

In all LCAs, the largest contribution to the total value in materials A, B and C comes from lithium, up to 80.00% in EI-99 for materials B and C. For material D, it comes from the element fluorine, at around 90.00% in three LCAs. Moreover, the second contribution element for material A is cobalt, at 38.78% in EI-99 and 48.67% in IMPACT 2002+. For materials B and C it is phosphorus, at about 20.00%. Other elements made just a small contribution to the total values, less than 5.00%. This result is consistent with Yang's [44] research that there is significant waste of valuable metallic resources in LIBs and the environmental load of lithium consumption is the largest among all elements. Furthermore, except for the contribution of lithium, phosphorus plays an important role in the environmental sustainability potential of materials B and C. For material A, the influence of cobalt also cannot be ignored.

### *3.5. The Methodologies' Influence to Environmental Sustainability Assessment*

Based on the discussion at the above four levels, we find that these four cathodes have different environmental sustainability potential due to the different LCA we use.

(1) Different LCAs have little impact on the environmental sustainability assessment of the total, endpoint, and element levels of the four cathodes. The environmental sustainability of $FeF_3(H_2O)_3/C$ is better than $LiFe_{0.98}Mn_{0.02}PO_4/C$, which is better than $LiFePO_4/C$ and $LiCoO_2/C$.

(2) The environmental sustainability of the four cathode types is mainly affected by human health. Lithium is the largest contributor to the environmental load of the first three market cathodes. However, for $FeF_3(H_2O)_3/C$, fluorine is the largest contributor to its environmental load.

(3) At the midpoint level, the four cathodes show different environmental sustainability in concrete indicators due to the different LCAs we chose. The mineral resource consumption of $FeF_3(H_2O)_3/C$, the metal resource consumption of $LiFe_{0.98}Mn_{0.02}PO_4/C$, high non-renewable resource consumption of $LiFePO_4/C$, and the toxicity and land issues of$LiCoO_2/C$ have seriously affected their environmental sustainability. Furthermore, the cobalt in $LiCoO_2/C$, because of its ecosystem quality, and the phosphorus in $LiFePO_4/C$ and $LiFe_{0.98}Mn_{0.02}PO_4/C$ also obviously impair the environmental sustainability at the midpoint level.

## 4. Conclusions

Different LCAs show different quantitative results in these four cathode materials. Qualitative assessments of these three LCAs is similar, both in terms of contributing elements and the total value. At the endpoint level, except for the resource consumption for $LiFe_{0.98}Mn_{0.02}PO_4/C$ and $LiFePO_4/C$ in ReCiPe, the ranking of other indicators' values is consistent with the total values. Four cathode models are ranked in descending order of environmental sustainability potential: $FeF_3(H_2O)_3/C$, $LiFe_{0.98}Mn_{0.02}PO_4/C$, $LiFePO_4/C$, and $LiCoO_2/C$. At the midpoint level, most indicators are consistent with the ranking. However, the most serious problem is determined differently based on different methodologies.

**Supplementary Materials:** The following are available online at www.mdpi.com/2227-9717/7/2/83/s1, Table S1: Calculation for mid point indicators in IMPACT 2002+, Table S2: Calculation for midpoint indicators in EI-99, Table S3: Calculation for midpoint indicators in ReCiPe.

**Author Contributions:** Conceptualization, L.W. and H.W.; Methodology, L.W. and H.W.; Software, Y.Y.; Validation, Y.Y. and K.H.; Formal Analysis, H.W. and Y.Y.; Investigation, L.W. and H.W.; Resources, Y.Y.; Writing—Original Draft Preparation, L.W.; Writing—Review and Editing, L.W. and H.W.; Visualization, L.W. and H.W.; Supervision, Y.Y. and K.H.; Project Administration, Y.Y. and K.H.; Funding Acquisition, Y.Y. and K.H.

**Funding:** This research was funded by: The Fundamental Research Funds for the Central Universities of China (No. 2018BLCB-05), the National Natural Science Foundation of China (No. 51474033), and Beijing Natural Science Foundation (9172012).

**Conflicts of Interest:** The authors declare no conflict of interest.

## References

1. Swart, P.; Dewulf, J.; Biernaux, A. Resource demand for the production of different cathode materials for lithium ion batteries. *J. Clean. Prod.* **2014**, *84*, 391–399. [CrossRef]
2. Zhu, X.; Lin, T.; Manning, E.; Zhang, Y.; Yu, M.; Zuo, B.; Wang, L. Recent advances on Fe- and Mn-based cathode materials for lithium and sodium ion batteries. *J. Nanoparticle Res.* **2018**, *20*, 160. [CrossRef]
3. Chakraborty, S.; Banerjee, A.; Watcharatharapong, T.; Araujo, R.B.; Ahuja, R. Current computational trends in polyanionic cathode materials for Li and Na batteries. *J. Phys. Condens. Matter: Inst. of Phys. J.* **2018**, *30*, 283003. [CrossRef] [PubMed]
4. Jeong, G.; Kim, Y.-U.; Kim, H.; Kim, Y.-J.; Sohn, H.-J. Prospective materials and applications for li secondary batteries. *Energy Environ. Sci.* **2011**, *4*, 1986. [CrossRef]
5. Cerdas, F.; Titscher, P.; Bognar, N.; Schmuch, R.; Winter, M.; Kwade, A.; Herrmann, C. Exploring the effect of increased energy density on the environmental impacts of traction batteries: A comparison of energy optimized lithium-ion and lithium-sulfur batteries for mobility applications. *Energies* **2018**, *11*, 150. [CrossRef]

6. Andersson, K.; Eide, M.H. The feasibility of including sustainability in lca for product development. *J. Clean. Prod.* **1998**, *6*, 289–298. [CrossRef]
7. Peters, J.F.; Baumann, M.; Zimmermann, B.; Braun, J.; Weil, M. The environmental impact of li-ion batteries and the role of key parameters—A review. *Renew. Sustain. Energy Rev.* **2017**, *67*, 491–506. [CrossRef]
8. Peters, J.F.; Weil, M. Providing a common base for life cycle assessments of Li-ion batteries. *J. Clean. Prod.* **2018**, *171*, 704–713. [CrossRef]
9. Li, L.; Dunn, J.B.; Zhang, X.X.; Gaines, L.; Chen, R.J.; Wu, F.; Amine, K. Recovery of metals from spent lithium-ion batteries with organic acids as leaching reagents and environmental assessment. *J. Power Sources* **2013**, *233*, 180–189. [CrossRef]
10. Yuan, L.X.; Wang, Z.H.; Zhang, W.X.; Hu, X.L.; Chen, J.T.; Huang, Y.H.; Goodenough, J.B. Development and challenges of lifepo4 cathode material for lithium-ion batteries. *Energy Environ. Sci.* **2011**, *4*, 269–284. [CrossRef]
11. Lin, Y.; Zeng, B.; Lin, Y.; Li, X.; Zhao, G.; Zhou, T.; Lai, H.; Huang, Z. Electrochemical properties of carbon-coated lifepo4 and $LiFe_{0.98}Mn_{0.02}PO_4$ cathode materials synthesized by solid-state reaction. *Rare Met.* **2012**, *31*, 145–149. [CrossRef]
12. Zeng, L.; Gong, Q.; Liao, X.; He, L.; He, Y.; Ma, Z. Enhanced low-temperature performance of slight Mn-substituted lifepo4/c cathode for lithium ion batteries. *Chin. Sci. Bull.* **2011**, *56*, 1262–1266. [CrossRef]
13. Togo, M.; Nakahira, A. Structure refinement of mn-substituted $LiMn_xFe_{1-x}PO_4$. *Mater. Sci. Appl.* **2018**, *09*, 542–553.
14. Wu, C.; Li, X.X.; Wu, F.; Bai, Y.; Chen, M.Z.; Zhong, Y. Composite $FeF_3{\bullet}3H_2O$/C cathode material for lithium ion battery. *Adv. Mater. Res.* **2011**, *391–392*, 1090–1094. [CrossRef]
15. Ludwig, J.; Nilges, T. Recent progress and developments in lithium cobalt phosphate chemistry- syntheses, polymorphism and properties. *J. Power Sources* **2018**, *382*, 101–115. [CrossRef]
16. Pfleging, W. A review of laser electrode processing for development and manufacturing of lithium-ion batteries. *Nanophotonics* **2018**, *7*, 549–573. [CrossRef]
17. Winslow, K.M.; Laux, S.J.; Townsend, T.G. A review on the growing concern and potential management strategies of waste lithium-ion batteries. *Resour. Conserv. Recycl.* **2018**, *129*, 263–277. [CrossRef]
18. Wang, Y.; Yu, Y.; Huang, K.; Chen, B.; Deng, W.; Yao, Y. Quantifying the environmental impact of a Li-rich high-capacity cathode material in electric vehicles via life cycle assessment. *Environ. Sci. Pollut. Res. Int.* **2017**, *24*, 1251–1260. [CrossRef]
19. Wang, C.; Chen, B.; Yu, Y.; Wang, Y.; Zhang, W. Carbon footprint analysis of lithium ion secondary battery industry: Two case studies from China. *J. Clean. Prod.* **2017**, *163*, 241–251. [CrossRef]
20. Liang, Y.; Su, J.; Xi, B.; Yu, Y.; Ji, D.; Sun, Y.; Cui, C.; Zhu, J. Life cycle assessment of lithium-ion batteries for greenhouse gas emissions. *Resour. Conserv. Recycl.* **2017**, *117*, 285–293. [CrossRef]
21. Deng, Y.; Li, J.; Li, T.; Zhang, J.; Yang, F.; Yuan, C. Life cycle assessment of high capacity molybdenum disulfide lithium-ion battery for electric vehicles. *Energy* **2017**, *123*, 77–88. [CrossRef]
22. Zackrisson, M.; Fransson, K.; Hildenbrand, J.; Lampic, G.; O'Dwyer, C. Life cycle assessment of lithium-air battery cells. *J. Clean. Prod.* **2016**, *135*, 299–311. [CrossRef]
23. Gong, Y.; Yu, Y.; Huang, K.; Hu, J.; Li, C. Evaluation of lithium-ion batteries through the simultaneous consideration of environmental, economic and electrochemical performance indicators. *J. Clean. Prod.* **2018**, *170*, 915–923. [CrossRef]
24. Xu, Y.J.; Huang, K.; Yu, Y.J.; Wang, X.M. Changes in water footprint of crop production in beijing from 1978 to 2012: A logarithmic mean divisia index decomposition analysis. *J. Clean. Prod.* **2015**, *87*, 180–187. [CrossRef]
25. Jolliet, O.; Margni, M.; Charles, R.; Humbert, S.; Payet, J.; Rebitzer, G.; Rosenbaum, R. Impact 2002+: A new life cycle impact assessment methodology. *Int. J. Life Cycle Assess.* **2003**, *8*, 324–330. [CrossRef]
26. Audenaert, A.; De Cleyn, S.H.; Buyle, M. Lca of low-energy flats using the eco-indicator 99 method: Impact of insulation materials. *Energy Build.* **2012**, *47*, 68–73. [CrossRef]
27. Lamnatou, C.; Motte, F.; Notton, G.; Chemisana, D.; Cristofari, C. Building-integrated solar thermal system with/without phase change material: Life cycle assessment based on recipe, usetox and ecological footprint. *J. Clean. Prod.* **2018**, *193*, 672–683. [CrossRef]
28. Henclik, A.; Bajdur, W.M. Application of selected methods of life cycle assessment to judgment of environmental hazard of production process of flocculant synthesized from waste phenol-formaldehyde resin. *Rocz. Ochr. Sr.* **2011**, *13*, 1809–1822.

29. Lamnatou, C.; Baig, H.; Chemisana, D.; Mallick, T.K. Environmental assessment of a building-integrated linear dielectric-based concentrating photovoltaic according to multiple life-cycle indicators. *J. Clean. Prod.* **2016**, *131*, 773–784. [CrossRef]
30. Cavalett, O.; Chagas, M.F.; Seabra, J.E.A.; Bonomi, A. Comparative lca of ethanol versus gasoline in Brazil using different lcia methods. *Int. J. Life Cycle Assess.* **2012**, *18*, 647–658. [CrossRef]
31. Owsianiak, M.; Laurent, A.; Bjørn, A.; Hauschild, M.Z. Impact 2002+, recipe 2008 and ilcd's recommended practice for characterization modelling in life cycle impact assessment: A case study-based comparison. *Int. J. Life Cycle Assess.* **2014**, *19*, 1007–1021. [CrossRef]
32. Starostka-Patyk, M. New products design decision making support by simapro software on the base of defective products management. *Procedia Comput. Sci.* **2015**, *65*, 1066–1074. [CrossRef]
33. Herrmann, I.T.; Moltesen, A. Does it matter which life cycle assessment (lca) tool you choose? A comparative assessment of simapro and gabi. *J. Clean. Prod.* **2015**, *86*, 163–169. [CrossRef]
34. Wanger, T.C. The lithium future-resources, recycling, and the environment. *Conserv. Lett.* **2011**, *4*, 202–206. [CrossRef]
35. Zhang, Y.; Huang, K.; Yu, Y.J.; Yang, B.B. Mapping of water footprint research: A bibliometric analysis during 2006–2015. *J. Clean. Prod.* **2017**, *149*, 70–79. [CrossRef]
36. Koellner, T. Land use in product life cycles and its consequences for ecosystem quality. *Int. J. Life Cycle Assess.* **2002**, *7*, 130–130. [CrossRef]
37. Klinglmair, M.; Sala, S.; Brandão, M. Assessing resource depletion in lca: A review of methods and methodological issues. *Int. J. Life Cycle Assess.* **2013**, *19*, 580–592. [CrossRef]
38. Li, W.; Song, B.; Manthiram, A. High-voltage positive electrode materials for lithium-ion batteries. *Chem. Soci. Rev.* **2017**, *46*, 3006–3059. [CrossRef]
39. Branco, H.; Castro, R.; Setas Lopes, A. Battery energy storage systems as a way to integrate renewable energy in small isolated power systems. *Energy Sustain. Dev.* **2018**, *43*, 90–99. [CrossRef]
40. Li, Y.; Yang, J.; Song, J. Design structure model and renewable energy technology for rechargeable battery towards greener and more sustainable electric vehicle. *Renew. Sustain. Energy Rev.* **2017**, *74*, 19–25. [CrossRef]
41. Peng, C.; Hamuyuni, J.; Wilson, B.P.; Lundstrom, M. Selective reductive leaching of cobalt and lithium from industrially crushed waste Li-ion batteries in sulfuric acid system. *Waste Manag.* **2018**, *76*, 582–590. [CrossRef] [PubMed]
42. Xiang, W.; Zhong, Y.; Tang, Y.; Shen, H.; Wang, E.; Liu, H.; Zhong, B.; Guo, X. Improving the electrochemical kinetics of lithium manganese phosphate via co-substitution with iron and cobalt. *J. Alloys Compd.* **2015**, *635*, 180–187. [CrossRef]
43. Di Lecce, D.; Brutti, S.; Panero, S.; Hassoun, J. A new sn-c/life0.1co0.9po4 full lithium-ion cell with ionic liquid-based electrolyte. *Mater. Lett.* **2015**, *139*, 329–332. [CrossRef]
44. Yang, Y.; Meng, X.; Cao, H.; Lin, X.; Liu, C.; Sun, Y.; Zhang, Y.; Sun, Z. Selective recovery of lithium from spent lithium iron phosphate batteries: A sustainable process. *Green Chem.* **2018**, *20*, 3121–3133. [CrossRef]

MDPI
St. Alban-Anlage 66
4052 Basel
Switzerland
Tel. +41 61 683 77 34
Fax +41 61 302 89 18
www.mdpi.com

*Processes* Editorial Office
E-mail: processes@mdpi.com
www.mdpi.com/journal/processes

www.ingramcontent.com/pod-product-compliance
Lightning Source LLC
LaVergne TN
LVHW070638170726
843515LV00004B/275

* 9 7 8 3 0 3 6 5 0 5 7 4 9 *